THE WINDOW GLASS MAKERS OF ST. HELENS

THE WINDOW GLASS MAKERS OF ST. HELENS

A record of the furnaces, ancilliary plant, the tools and the equipment; the skills of the Brothers and of the men and the women, who made and developed the art of making window glass from 1826 to 1952.

R. A. Parkin, BSc, CEng,
FIMechE, FIEE

Published by
Society of Glass Technology
2024

The Window Glass Makers of St. Helens by R. A. Parkin, BSc, CEng, FIMechE, FIEE

The objects of the Society of Glass Technology are to encourage and advance the study of the history, art, science, design, manufacture, after treatment, distribution and end use of glass of any and every kind. These aims are furthered by meetings, publications, the maintenance of a library and the promotion of association with other interested persons and organisations.

Society of Glass Technology
9 Churchill Way
Chapeltown
Sheffield S35 2PY, UK
Tel +44(0)114 263 4455
Email info@sgt.org
Web http://www.sgt.org

The Society of Glass Technology is a registered charity no. 237438.

ISBN 978-0-900682-97-1

[Revised figures, minor text corrections]

CONTENTS

Introduction

This is a record of window glass making by the Company of Pilkington Brothers, at Grove Street, St. Helens from 1826 to 1952.

St. Helens is in England in the South of Lancashire at what was in the old days the junction of the roads from Liverpool to Bolton and from Warrington to Ormskirk. It grew up from 1700 onwards, as a product of the industrial revolution when it was found to have coal and sand in its immediate vicinity with salt nearby. The construction of a canal and then a railway helped to found glass and chemical industries, copper works and foundries, all of which grew steadily through the 1800s.

There are three periods in history of glass making at Grove Street, the first being what I will call the "Manual Age" from about the 17th. Century to 1930, the second being the "Mechanical Age" from 1930 to 1958 and then from 1958 the "Technological Age", epitomised by the making of glass by the float process.

Manual Age	1600 to 1930
Mechanical Age	1930 to 1958
Technological Age	1958 to 2000

The "Manual Age" is the period, when everything was made by hand by employing many thousands of men, women, boys and girls. The "Mechanical Age" that followed represented the introduction of machines for the making and handling of window glass by a labour force that was still substantial, but much reduced. In the "Technological Age" glass was and still is being produced, virtually untouched by human hand, under the technological control of a small labour force.

The role of the Pilkington Brothers has long been recognised but it takes more than a hierarchy to establish the foundations of an industry that has become an influence world wide. The three "Ms", "Men, Materials, and Machines" are the components that have to be brought together by management and in this record it is the men and women to whom we pay tribute. They, the people of St. Helens who can call themselves glass makers, gatherers, blowers, teazers, producer men,

Figure 1. ***A picture of St. Helens dated 1834 which shows the cone dominating the landscape. In front of it are the warehouses and the chemical factories on the side of the canal***

splitters, cutters, carriers, supplemented by the claymakers, masons, smiths, sand getters, and later the chemists, engineers and designers, all of whom inherited that unique spirit and character that was inherent in that little township.

The record is based on what has been ascertained by research into the archives of the company. It is also based on what has been told by those few who could remember something of the very early days of glass making and what was revealed when the foundation of one of the early glass making tanks was excavated on the Jubilee site at Sheet Works in St. Helens. It was this excavation that drew attention to the possibility that the company was in danger of losing its heritage, particularly when it was found that all the early engineering drawings had been destroyed and that many other records had been lost.

Glass making was a closely guarded secret and little was recorded in the period from 1826 to 1896. Except for drawings made by Siemens and a few others no drawings of the plant or furnaces were made by Pilkington until after 1900. Fortunately, there were a few drawings available showing site layouts and some artist impressions of the works.

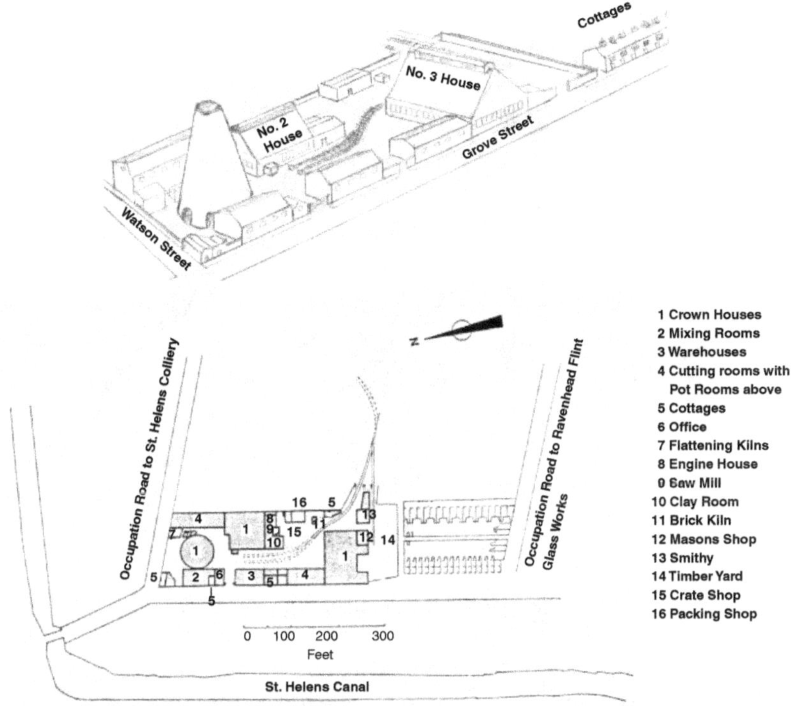

Figure 2. *The Pilkington Glass Factory in 1840*

Research showed that from the beginning, glass making was a progressive evolution of glass making techniques and furnace design, brought about as and when new materials became available and the demand for window glass increased. It also revealed the sheer skill and ingenuity of the glass makers of those early times.

Background.

The record is wholly about the sheet and rolled glass making factory known as Sheet Works, located in St. Helens at Grove Street only a short distance from the town centre. It was the birth place of the Company of Pilkington Brothers when it was originally founded as the St. Helens Glass Company in 1826 who started glass making under the Old Cone or No. 1 House, near to the banks of the St. Helens canal.

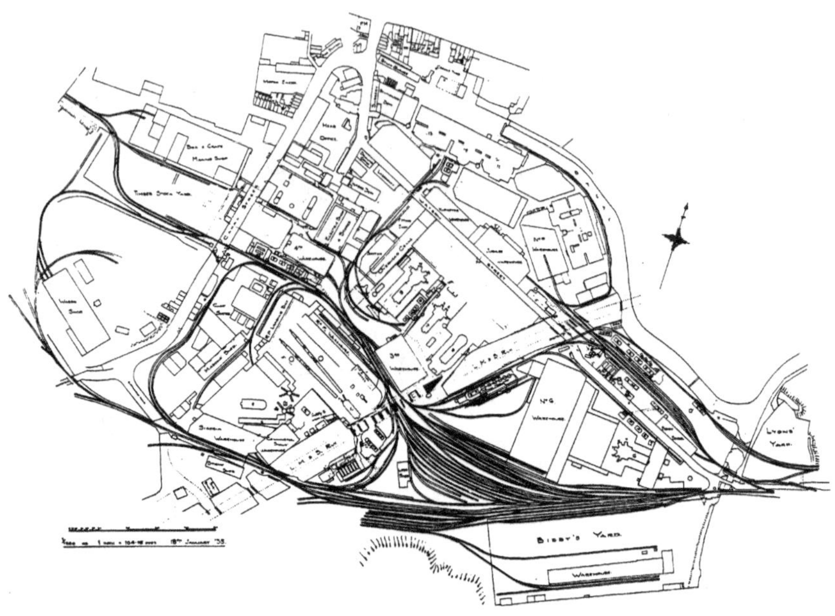

Figure 3. ***Plan of works in 1930***

The cone covering the glass furnace is clearly shown in the picture of St. Helens in 1834 (Fig. 1) and the drawing of the works as it existed in 1840 (Fig. 2). The outline plan of the works as it was in 1840 shows the cone, the canal, the roads and the railway. By 1879 the works had expanded to make room for more furnaces. In 1901 the canal was filled in, Grove Street was built over, and more land bought, so that by the 1930s the works then occupied a huge area holding three and still later four 1000 to 1200 ton sheet glass melting furnaces, and two 350–500 ton rolled plate glass melting furnaces. It made window glass, horticultural glass, and glass for the motor trade on what were called the Flat Drawn Tanks. It made patterned and wired glasses on the Rolled Plate Tanks.

The factory covered a huge area (120 acres), Fig. 3. It was almost completely self contained with its own timber yard; crate and box making shops; warehouses for cutting up the glass and packing it for shipment; gas making plant for firing the furnaces; fitting and electrical shops; boilers and compressed air plants; railway sidings and locomotives; clay sheds and masons sheds for making the furnace refractories. Timber was brought into the factory from Norway; sand from the local sandfields; non-caking coal from the pits in the Midlands; chemicals from Northwich, and limestone from Wales. All told there were some 3000–4000 foremen and workers managed by a relatively small number of staff. There was a small wage's department, one employment man, no technical staff, numerous small offices dealing with orders, time study, stores, diamond setting, and a shop called wire weaving which made the wire mesh for the wired glasses. It was a veritable hive of activity which does not compare with modern plants in this technological age where the proportion of managers and technicians to workers is completely reversed.

CHAPTER 1.

Glass Making in the 17th & 18th Centuries

The only interest in the early history of glass making has been to obtain some knowledge of glass making before 1826 in order to establish the state of the art before the founding of the Pilkington Glass Works at Grove Street, St. Helens.

Prior to the 17th Century the recorded history of the various means for melting glass was very vague. Even as late as the 18th Century there were old out dated furnaces still in existence and in operation in France. This and the few historical documents left to us, has not made it easy to identify the evolution of glass making, from an oven on the forest floor to the 1200 ton sheet and plate glass melting furnace of the 1930s. The main source of information has been the book "Le Verre" by Henrivaux of St. Gobain (1897) with references in that book to Theophilus and Agricola and the book on glass making "Guide du Verrier" by G. Bontemp written in 1868.

The Bee-hive Glass Melting Furnace

Somewhere in the 17th Century and possibly before that there was the growth of the 'bee-hive' furnace, for fritting, melting and fining with an oven for annealing. Firing was mainly by wood, and even peat was mentioned. This type of furnace was probably made possible when clay bricks came into use and is recorded in some detail by Henrivaux, Fig. 4. It was used for making small glasses, jugs, vases, and possibly some flat glass from small blown cylinders, which were cut with shears and flattened on either marble or clay blocks.

The English Cone or Glasshouse (The Verrerie Anglais)

There were indications that as the size of the furnaces increased, the atmosphere around them became more and more untenable, not only for the workers but for draughts of cold air which could cause breakage. The need to remove the smoke and soot and to create a stable atmosphere was solved by building an open-ended cone around and

Figure 4. *The beehive furnace*
An illustration of a glass making furnace taken from the book by Henrivaux

over the furnace. The cone increased the draught through the grate without killing the flush from the furnace, at the same time it kept the atmosphere around the furnace at a steady working temperature. In England the historical notes made by T. C. Barker were very helpful in indicating that the cone type of glass furnace construction was in use by 1700 and an old print dated 1743 showed a cone in the village of Prescot. It would seem that the use of a cone was an almost entirely English development. Henrivaux only mentions the "English Glasshouse" in passing in his book and he gives little information on what was in use in France between 1700 and 1850.

The continent appeared to have been slower in developing more advanced techniques in window glass making, although they advanced rapidly in the casting and making of plate glass.

The early Verrerie Anglais consisted of a cone built around a coal burning furnace possibly developed from the earlier type of beehive furnace, Fig. 5. The design and concept of this furnace was most ingenious for all the operations for glass melting and making were self contained, leaving the cutting, packing and dispatch as a separate operation in a warehouse building.

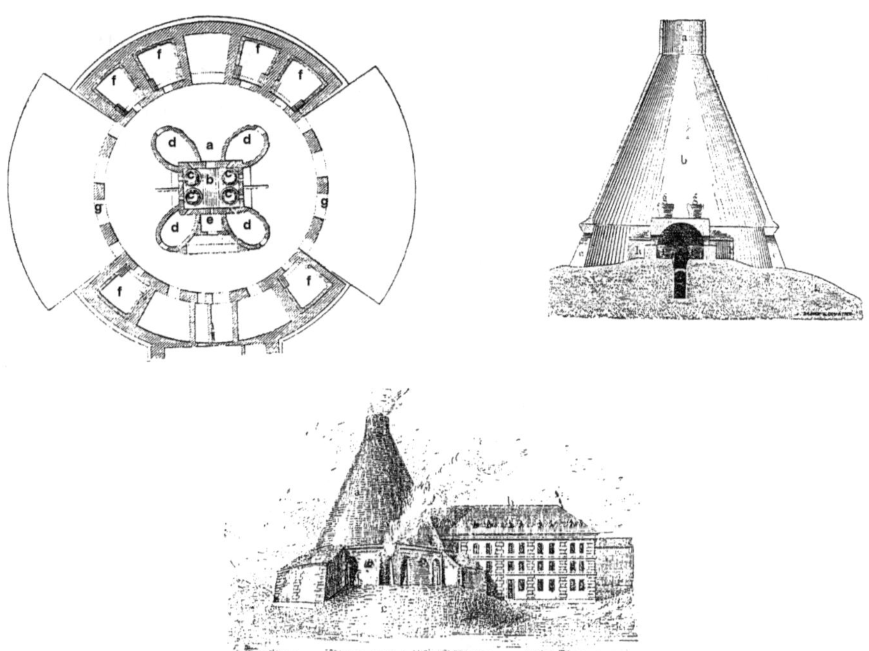

Figure 5. ***The Verrerie Anglais***

This is an English Glass House with a pot melting furnace inside the cone, and outside the cone, the annealing kilns and the warehouse where the glass tables were cut into window panes and quarries ready for dispatch. In the centre of the cone is the pot melting furnace, (b)(c) is the grate and the pots, (d) the pot arches, (e) the fritting kiln, (f) the annealing kilns, (g) the doorway into the cone, as well as pots drying on the top of the furnace

The heart of the Verrerie was a coal fired central furnace fed with air from an underground flue, the cone providing additional draught. The frit was melted in pots. Once melted, and refined for gathering, the final product was annealed ready for dispatch to the warehouse.

The batch consisting of sand, limestone, and saltcake was mixed by hand and calcined into frit in the calcining chamber. The pots that had been made and dried over quite a long period were fired in the pot chamber. The furnace itself was built with fireclay bricks, fired and brought up to temperature, with the pots in place. The pots were then glazed with cullet and filled with frit. The filling holes were made up leaving holes for gathering and blowing. The furnace was then raised to full heat to melt the frit. Once melted the glass in the pots was left to fine over a period of some hours. Once the fining had been completed the temperature was lowered for gathering and making glass.

The Pent Roofed Glass Houses.

In 1790, there was a visit to the Crown and Broad Glass Houses in Newcastle as recorded in William Baynton's Diary October 1790. Here crown glass was made in a square house with a wooden pent roof supported by wooded pillars, Fig. 6. In the *Illustrated Itinerary of the County of Lancaster* 1842 there was an illustration of a pent roofed glass house with smoke pouring out of the ventilators in the apex. It was this that gave the clue to the construction of what could be a much cheaper way of venting the gases and at the same time provid-

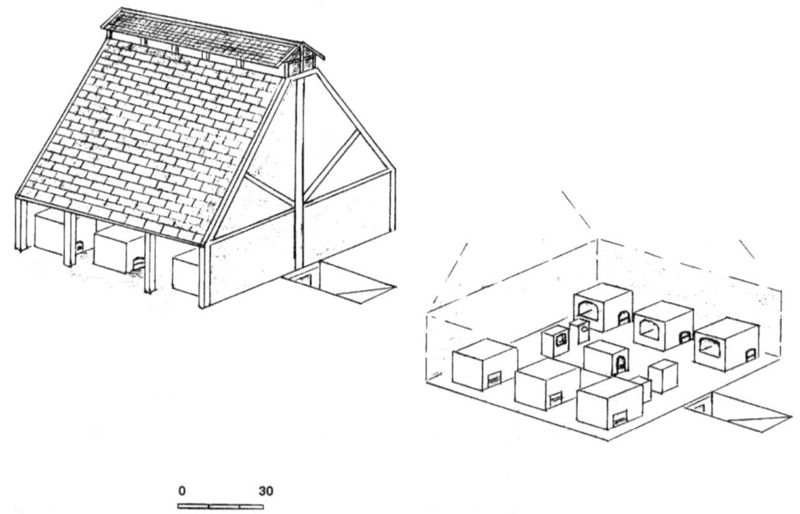

Figure 6. *The Newcastle Glass House 1790*
An interpretation of the description of the glass house as given in Baynton's Diary showing the building to be at least 60 feet high with the cave underneath and the possible layout of the furnace, flashing, bottoming and annealing kilns
Scale approx. 1:300

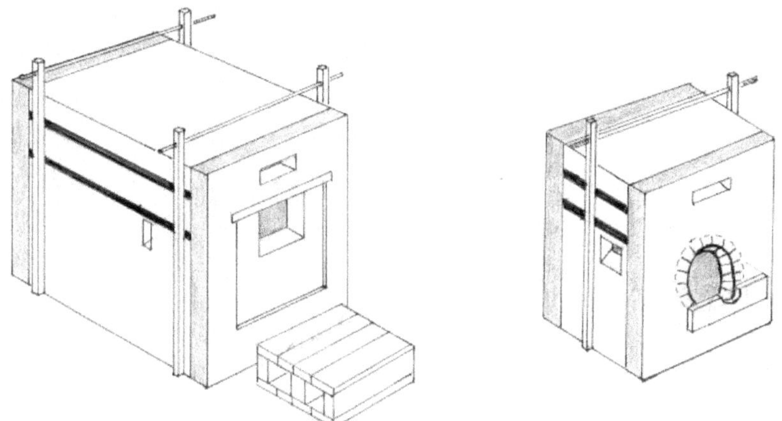

Figure 7. ***The Newcastle Great Furnace and the flashing furnace***
Scale approx. 1:30

ing the furnace draught. Inside the house was the Great Furnace for melting the batch into molten glass in the pots, and flashing furnaces for blowing and flashing the glass pieces, with six annealing kilns located down the sides. The furnaces and the kilns were partly dimensioned. Although the furnaces were fired from below in a cave it is likely that the annealing kilns were fired at floor level, for they only needed to be heated to a lower temperature.

The Great Furnace

The Great Furnace (Figs 7 & 8) could only accommodate two small pots, holding about 900 to 1000 lb of glass. Discarded pots were pushed off the siege into the cave and replacement pots slid in through the teazer hole on a sheet of iron. The top of the Great Furnace was squared off with common brick to serve as a pot scaffold. The pontils were heated at two small holes 5″ by 1″ at opposite corners of the Great Furnace, and the pipes were heated at the Flashing Furnace. The Great Furnaces lasted about two years. Founding took about 20 to 21 hours. There were no details about the fill but it would seem that a batch was about 4 cwt which they burned in about 4 hours. By burning, it was assumed that this was the time required to make it into frit.

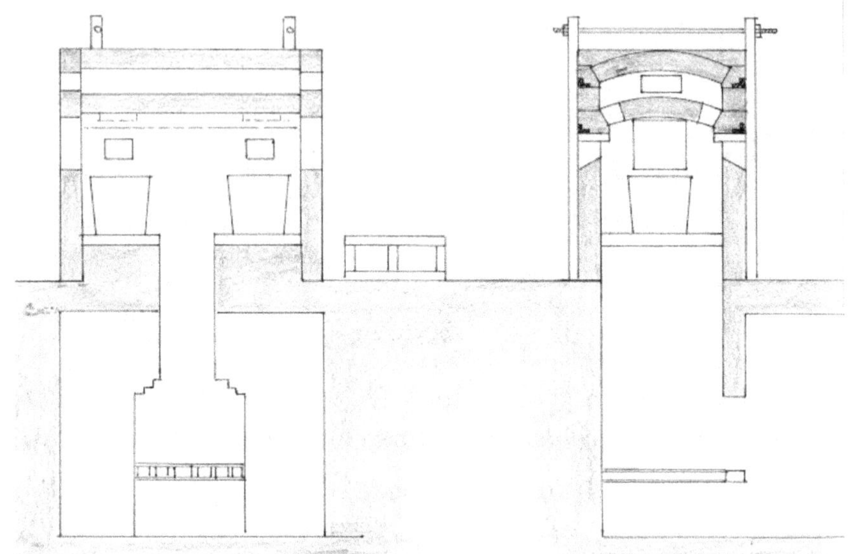

Figure 8. ***Section through the Newcastle Great Furnace showing the grate and the cave beneath, the reverbatory crown, and a gathering pat height of 2' 2"***
Scale approx. 1:30

CHAPTER 2.

Glass Making in St. Helens from 1826 to 1850

The St. Helens Crown Glass Company was started in 1826 by a group of influential people that included William and Richard Pilkington. It was based on the technical knowledge of a William Bell, who was described as a glass maker but not a specialist in crown glass making. He built a cone that must have been unbelievable large in its day, seemingly based on the design of a Verrerie Anglais. It was 66′ in diameter, 120′ high, with an opening at the top of 14′. There is a possibility that it contained a round reverbatory pot furnace of the type indicated by Barker in his book, Fig. 9. This pot furnace was about 20′ in diameter and it fitted in so well that it would give the gatherers and flashers

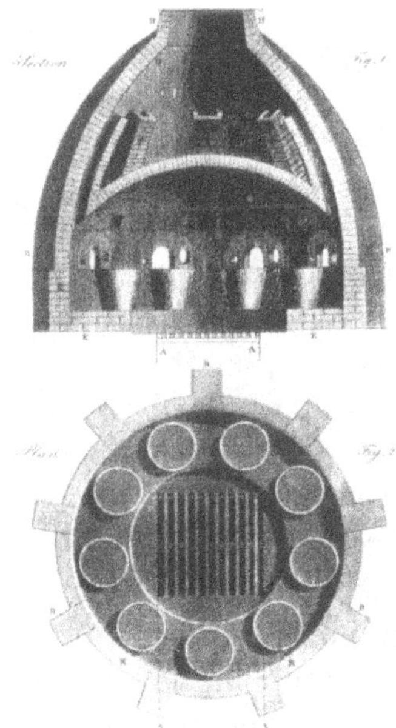

Figure 9. *The round reverbatory furnace*
This may be the type of furnace that was built and worked inside the cone when production started in 1826/7 complete with teazing holes as well as gathering and flashing holes
(from Pantalogia, a new Cyclopaedia, *1813)*

enough room to work. There were no nosing, bottoming and flashing furnaces for all the operations of gathering, nosing, bottoming and flashing were carried out over the pots in the round furnace.

It was built on the corner of Grove St. and Watson St. and at 120′ in height it dwarfed everything in sight. The combustion air flue or tunnel underneath the furnace went right through from Grove Street into the yard beyond and it was remembered by some of the workmen who worked in the 1920s as they walked up Grove St. to check in at the Old Lodge.

The round reverbatory furnace was coal fired. It melted glass in possibly 8–10 pots with up to 4 cwt, maybe 5 cwt, of glass in each pot. It was operated and worked in much the same way as its forerunner the Verrerie Anglais, except that it was used mainly for making crown glass with some broad or spread glass.

The Rectangular Pot Melting Furnace

Making glass in a round pot melter was very tedious and slow. It was slow because the holes over the pots in the furnace were not only used for gathering but for forming and flashing. The plan of the works in 1840 shows that no more cones had been built, instead they had installed by that time two square glass houses. The only explanation may be that under the influence of William Pilkington; and experience elsewhere, plus the need to increase production; they adopted the Newcastle style of glass house which with a pent roof, which acted the same way if providing the draught, was much cheaper to build than a cone and by using separate flashing, nosing and bottoming furnaces they could increase production. The new rectangular glass house, some 60′ to 70′ in height, was adopted and built in 1834. This was No. 2 House and it could have housed a small two or four pot furnace with all the separate blowing and flashing furnaces. The need for even greater production of glass was such that by 1836 No. 3 Glass House had to be built, see Fig. 2. The works plan shows that it covered a bigger area so we can presume that it was built on the same lines as No. 2 with a larger furnace using even more and possibly larger pots, each holding about 1000 lb of metal.

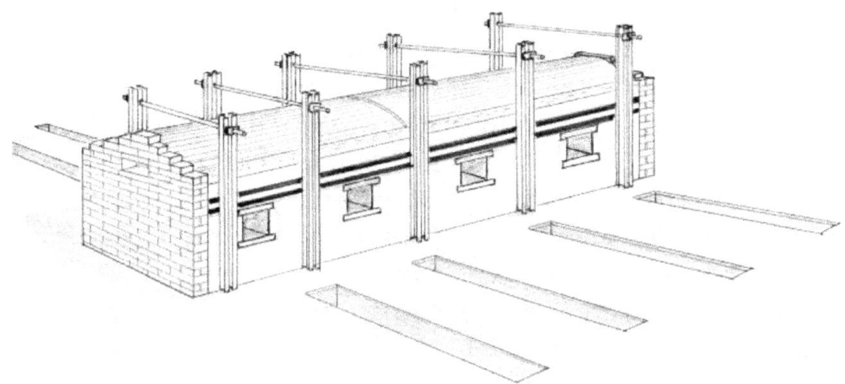

Figure 10. ***A reverbatory coal fired blowing furnace with offset blowing holes and swing pits***

Scale approx. 1:60

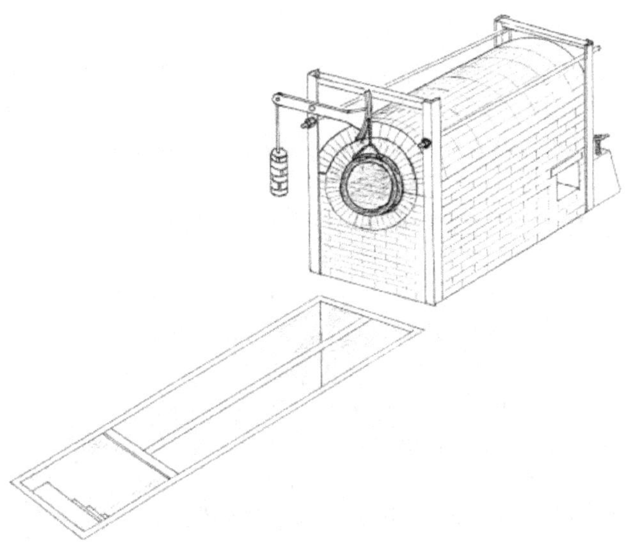

Figure 11. ***Glory Hole***

In later years, Glory Holes were built in place of large blowing furnaces to meet individual requirements on a tank making a variety of products, such as cells, shades, lamp glasses, and lagres. It was a small tubular furnace heated by a gas jet with a compressed air operated tweel. The swing pits were there to accommodate either hand or machine blowing and the steps down were for access for removing cullet

In the period from 1841 to 1850 the production of crown glass was being phased out and replaced with blown cylinder sheet glass. Also in this period Nos 4, 5, and 6 Glass Houses were built complete with Blowing Furnaces, needed for the blowing of the cylinders, Figs 10 & 11. There were two blowing furnaces for each pot furnace. For instance an 8 pot furnace with 8 gathering holes, needed 16 blowing holes, that is 8 blowing holes in each of two blowing furnaces. The Blowing Furnaces virtually replaced the Flashing Furnace. The Blowing Furnace unlike the Flashing Furnace needed stable temperature conditions around it for glass blowing so each one had a small cone built over it. It was a drawing of the plan of the works dated 1878 and the artists impression of the works in 1879 (Figs 12 & 13) that gave a clue to the fact that the blowing furnaces were built inside a small cone, 18 yards high, as compared to the Old Cone which was 120 feet high, in order to regulate the conditions for blowing.

Throughout the period 1834 to 1850 the coal fired pot melters were being steadily being increased in size to accommodate more and possibly larger pots, so that by the end of that time the company had pot furnaces working with 8–10 pots inside them.

The Construction, Operation and Firing of a Rectangular 8/10 Pot Melting Furnace

(see Newton's London Journal of Art. Richard Pilkington 1864 L 1405)

The rectangular pot melter complete with a reverbatory crown was made of fireclay bricks. The 8–10 pots were on a raised bed (siege) which extended the full length of the furnace. The grate was 4 feet shorter at the far end. The air was supplied from an underground flue with doors to control the draught. The history of the reverbatory crown as exemplified in the Round Pot Melter, was that it was introduced to stop soot and droppings coming off the crown and falling into the pots and to provide reflected heat. We have read that coal was becoming available from 1610 to 1650 for mining was very difficult owing to drainage problems. As and when these problems were overcome production in the downbrew mines and the collieries got better so that by 1750 coal became more freely available. Soot and droppings from the roof of a furnace contaminated the glass in

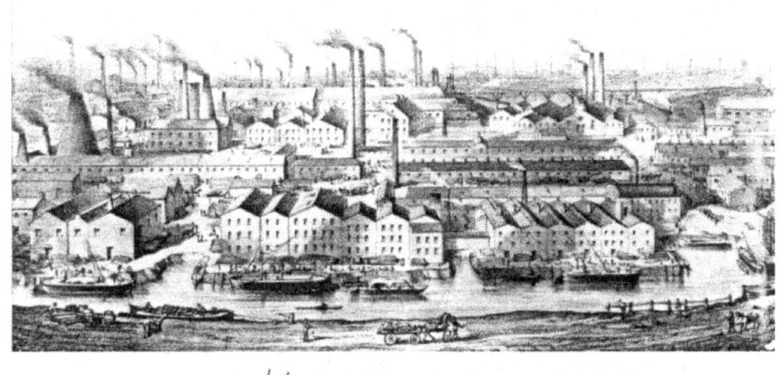

Figure 12. ***An artist's impression of Sheet Works with the canal in the foreground as it was in 1879. The "Old Cone" or No.1 House with the two small cones over the blowing furnaces are clearly shown. Below it is a rough copy of a works plan recently discovered in the Archives attached to the valuation sheets dated 1878. This is the first time that it has been possible to relate an impression with a contemporary plan***

the pots, giving rise to covered pots and the reverbatory furnace. Up to 1830 coal was just burnt on a grate within a confined space to melt the frit. Somewhere between 1840 and 1855 it was found that by pouring a trickle of water on to the grate an even better flame was produced. The explanation is that provided the temperature

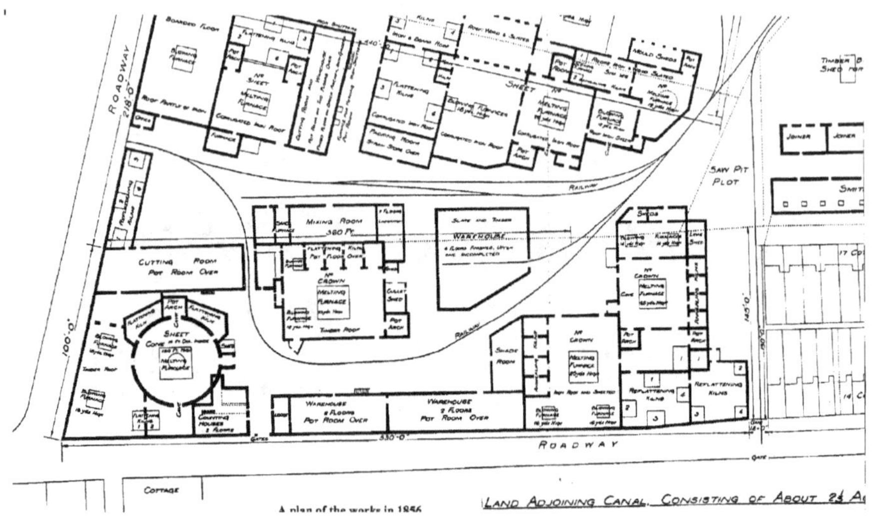

Figure 13. ***A plan of the works in 1856***

at the heart of the furnace was kept high enough, the water and the steam was broken down to combine with the carbon to form carbon monoxide and hydrogen. As time went on, and more was learned about deep firebeds, the use of slack as well as non-caking and non-bituminous coals, it must have been realised that it was the gases that were being given off that produced cleaner flames in the furnace. This was probably the forerunner of the producer gas concept and at the same time the role of a reverbatory crown changed from that of reducing contamination to that of providing reflected heat.

The pots and gathering rings were made of the best Stourbridge fireclay and left to dry for a year at 60°F and then left at 90°F until wanted, either already fired or hot from the pot arch. There was no mention of grog being used. They were 5′ in diameter overall at the top 4½′ diameter inside at the bottom and 4′ deep weighing 25 cwt. The weight of metal was 22 cwt; they cost £9 and lasted 8 weeks. The gathering ring was 18″ internal diameter and 2″ thick.

The illustration is of an 8 pot coal fired reverbatory pot melter of about 1850 containing standard 42″ pots using a standard pot height of 2′ 2″ above gathering floor level, Figs 14 & 15. Note that on every

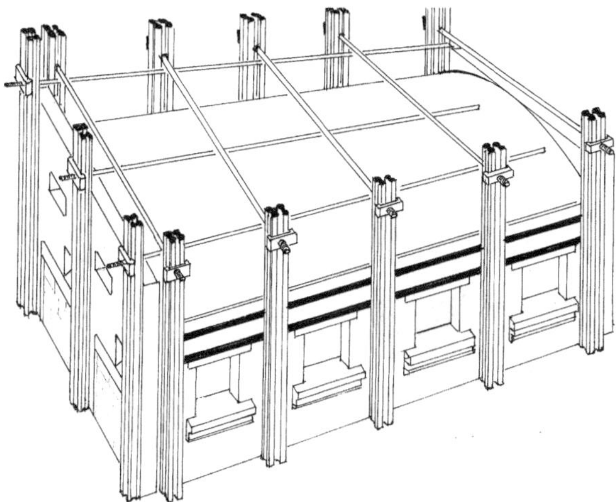

Figure 14. ***A Coal Fired 8 Pot Reverbatory Glass Melting Furnace with provision for end loading the pots on to the siege and side teazing and gathering holes***
Scale approx. 1:30

gathering or ladling hole in a furnace side the flat base of the hole is called the pat and the height of the pat for gathering was critical. The furnace was braced with railway metals and the thrust of the crown springers possibly taken on cast angle iron inserts (as illustrated in Siemens' drawing of a Centre Fired Gas Furnace). It was most likely that the end walls were built outside the ends of the crowns and that they held the pipe holes for warming the pipes. The furnace would be operated in such a way that by controlling the size of the vents in the end walls there would always be a flush at the gathering holes.

The pot melting furnace was brought up slowly to the working temperature and the red hot fired pots loaded through the ends and the ends closed. Cullet was then put into the pots and the pots glazed on the inside using a ladle. The next stage was to fill the pots with frit. This was done in four stages, four charges in each stage with an interval between each stage to allow the frit to melt. The reason for doing it in stages was the fact that each melt was accompanied by a heavy boil. When the last stage had been completed, the teazing holes were closed with a stopper and the glass in the pots fined for 8 hours.

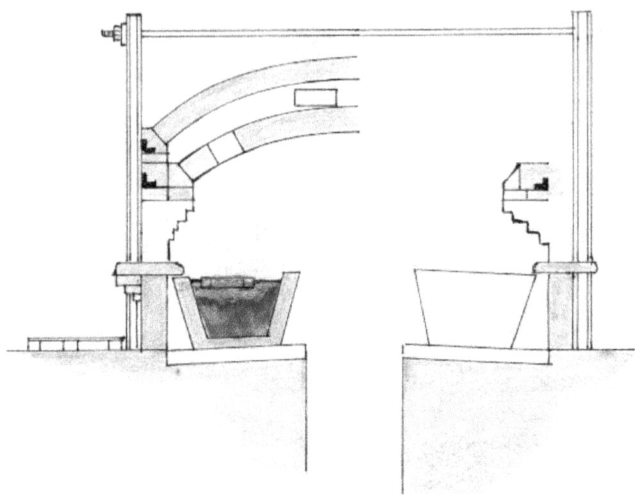

Figure 15. ***A Section through a Pot Furnace showing the reverbatory crown, teazing hole, pot and gathering ring using a pat height of 2' 2"***
Scale approx. 1:30

Finally, the pot was skimmed inside the ring using an iron rod with a chisel shaped end. The process was an extremely slow one. The pots had to be filled; the frit melted; the glass in the pot refined to release all the bubble; then the temperature lowered for the gatherers to gather the glass; form the bulb, attach the punty and spin the crown.

The gatherers worked until all the pots were emptied and the crowns or the tables as they were known placed in the annealing kilns, after which they all went home and only returned when the pots were ready for gathering once again. This was known as the 'found' and all told there were three to four founds, rarely five, in any one week. During their absence and whilst the pots were being refilled the annealing kilns were cooled over a period of 24 to 36 hours and after cooling, the crowns or tables were sent to the warehouse for cutting into window panes. To avoid the inroad of cold air, the coal fired annealing kilns did not have any chimneys, but the pot kilns did for they were fully sealed.

The routine was much the same for the production of hand blown sheet glass after the production of crown glass had been phased out.

CHAPTER 3.

The Making of Broad or Spread Glass and Crown Glass up to 1850

Broad or Spread Glass

In centuries long ago, glass makers could make flat window glass by either casting the molten glass from pots on to a flat surface or by gathering and blowing it in the form of a cylinder. Casting may not have worked very well for thickness or clarity, but glass gathered on a blow pipe, blown into a cylinder and flattened, was very much better. This was called broad or spread glass and the method of making it is illustrated in an old 18th Century illustration, Figs 16 & 17.

It was made by first of all by gathering molten glass on the end of a blowpipe and then blowing it into the form of a bulb. They then pierced the end and opened the bulb up with the tongs. Having done this the open end was sheared along its length and held on a disc so that the blowpipe could be cracked off. That end was then heated, opened up and sheared so that the cylinder could be flattened and made into a flat sheet of glass.

There are three ways of cutting glass. The first way was to shear it when it was still soft, the second way was to split it with hot iron when it had hardened, and the third way was the use of a diamond or cutting wheel when it was cold. Shearing was a somewhat primitive way of and no doubt it was succeeded by using a hot iron, a string of molten glass or hot wire with a drop of water. Diamonds were used as and when they became available to cut glass when it was cold.

Crown Glass

Broad or Spread glass had its limitations and in the 18th Century it was overtaken by the manufacture of glass by the crown method. Glass produced this way was much better for thickness and clarity. It is not known how crown glass was cut into squares when it was first produced. It may therefore be assumed that the cutting of glass by diamonds was available at the time. Diamonds according to Bontemp

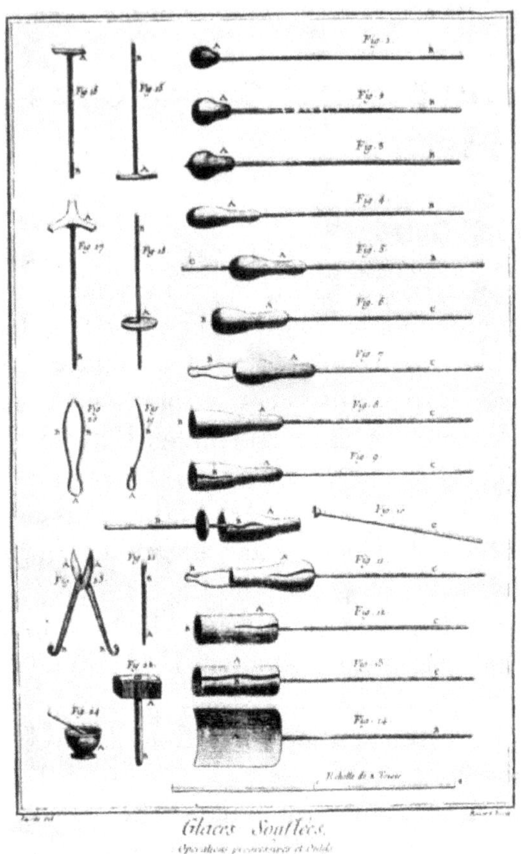

Figure 16. ***An 18th Century illustration of the making of broad or spread glass***

were in use in the 16th Century. Crown glass was certainly made by 1750 and predominantly so right up to 1826 when the company started to make glass in St. Helens.

In William Baynton's report of his visit to Newcastle in 1790 to see crown glass being made, he noted that: there the blowpipes were 6′ 8″ long with lips at the nose, the length of taper being 3″ to 4″; they gathered 9 to 9½ pounds of metal and quote "marver it down to make a thin nose, a strong shoulder and a little moyle; the blower then takes over and with the first heat blows it thin in the bullion and quite round." The pontils are 6′ 11″ long and thick at the end. If they gathered 9 to 9½ lb of metal it means that one pot would make 100 tables of crown glass.

The Ponty Stickers did not gather more than ¾″ long and were responsible for sticking and warming their own pieces. The Flashers carried off the pieces to flash out in the flashing furnace using a screen or shade with a square hole in it to observe when the piece was flashed. After that the piece was sheared off the pontil and placed in the annealing kiln. Their tables worked out at about 30″ in diameter for 24 oz glass. To blow a piece 48″ diameter they gathered 10 lb of metal and to get this the thickness would be about 2 mm.

Later, according to Henry Deacon's article "The Manufacture of Blown Window Glass" Feb. 1851, crown glass in the 1840s was made by the gatherer gathering on the end of his blowpipc sufficient metal to form a table of crown glass which was about 52″ in diameter and weighing 13 to 14 lb, Fig. 17. The blowpipe was about 6′ long but bear in mind that there were about three different lengths to reach into the pot as the level of the molten glass went down. He might have to make two or three gathers before he got the weight.

The glass or piece was rolled on the "marver" to shape it so that the point will ultimately form the bullion. The marver was made of polished cast iron and in the past a flat slab of marble.

The piece was then blown and handed on to the blower who reheated it in a small reverbatory furnace through what was called the Patteson Hole. It was swung downwards, marvered, and blown to form a bulb with a bullion at the point. It was then reheated in the Patteson Hole and blown against a bullion socket which cooled it and prevented it from expanding.

The blower then went to the Bottoming Furnace, which was much larger than the Patteson Hole, and there he reheated the piece and blew after two heats it into a globe. After a third heat the globe was spun and flattened out. At this point the punty was attached to the bullion socket. The punty was an iron rod with a ball of hot metal on the end. The blowpipe was then cracked off leaving an opening about 2½″ in diameter. This opening had very rough edges which was smoothed out and opened up at the "nose hole", a very small

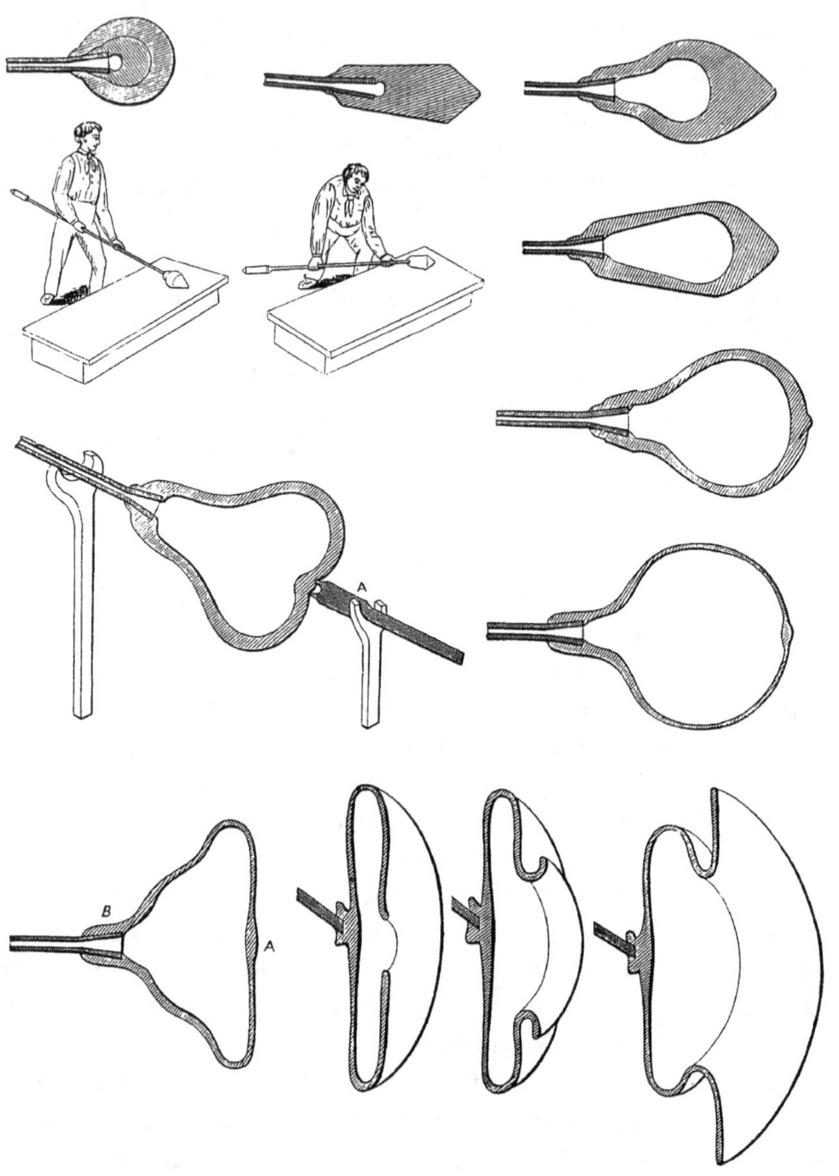

Figure 17. ***These drawings taken from Henry Deacon's paper on the manufacture of blown glass window glass show the sequence of operations from the gather to the shaping and finally the blowing and flashing of a table of crown glass***

but intensively heated reverbatory furnace.

The piece was then delivered to the "piece opener" who worked at the Flashing Furnace. The Flashing Furnace was another quite large reverbatory furnace, run at a high temperature with a large opening capable of accommodating the piece when it was flashed. The piece was heated in this furnace and rotated so that it gradually opened out and flashed into a circular plate. This plate was kept rotating by the carrier until it was stiff enough not to bend. It was then laid on a whimsy and the punty detached by a pair of shears. As soon as this was done the table was lifted using a horned fork and piled in the annealing kiln which could hold from 350, to 400 tables.

The men worked from 14 to 16 hours in what was called a journey without any interval. There were four workmen to a set, and as three only were engaged at any one time each man was able to rest for a short time. Normally they did four journeys per week and between 600 and 750 tables each journey. The glass was allowed to cool for 48 to 60 hours in the annealing kiln.

In 1826 when the factory at Grove Street, St. Helens first started, it made crown glass and just a little broad glass by the methods that have just been outlined. Crown glass compared to broad glass was quite good but it had its limitations. In cutting squares out of a round table, the cutting loss was so high that there were large amounts of waste glass called cullet. The maximum size of any one square was 34″ × 22″ and there were some variations in thickness.

Nearly all the glass made between 1826 and 1841 was crown glass, but by 1841 crown glass started to be phased out and replaced by hand blown cylinder glass using improved methods, and by 1874 the making of crown glass ceased altogether.

This hand blown glass known as sheet glass was a big improvement on crown glass for quality and thickness even though flattening marks were a drawback. The improvements were in the size of the cylinder, and the fact that 40″ × 30″ panes of glass could be produced with much less waste.

CHAPTER 4.

The Developments in Gas Firing and Pot Furnaces from 1850 to 1870

The Siemens Regenerative System

From 1850 to 1860 Siemens was working on the regeneration of the waste heat by preheating of the gas and air.

Instead of burning coal directly under and round the pots, the waste heat going into the atmosphere, coal gas and air was introduced into the furnace through burner ports on the in-going side, where it burnt and left the furnace through a set of ports on the outgoing side. On leaving the furnace the outgoing hot gases passed through and heated a lattice work of bricks. After say 20 to 30 minutes, the air and the gas was reversed, so that the incoming air and gas was now preheated by passing through the red hot brickwork, and the outgoing gases reheated the other side, thus giving rise to much higher temperatures and greater efficiency in the operation of the furnace. The lattice work of bricks were in what was called a regenerator, a separate one for the gas and another for the air. The burners were called stacks with separate upcasts and mid-feathers, for the gas and the air.

This meant that between 1850 and 1860 Siemens had developed the gas producer, probably before the regenerative system for a coal fired furnace in 1857. It would also be about this time that chimneys came into use, for it was not possible to have a chimney on a coal fired pot furnace because of the in-drawn air. Siemens was very fortunate in that the expansion of the preheated gas and air and the subsequent combustion so balanced the chimney pull that the furnace was always kept under pressure.

In 1862 in the book written by Bontemp there was a drawing of an inclined grate producer (Figs 18 & 19) with water being introduced down the gas offtake. Thus between 1860 and 1870 starting with the deep coal fired grate with water added, we had the progressive development and application of the inclined grate producer.

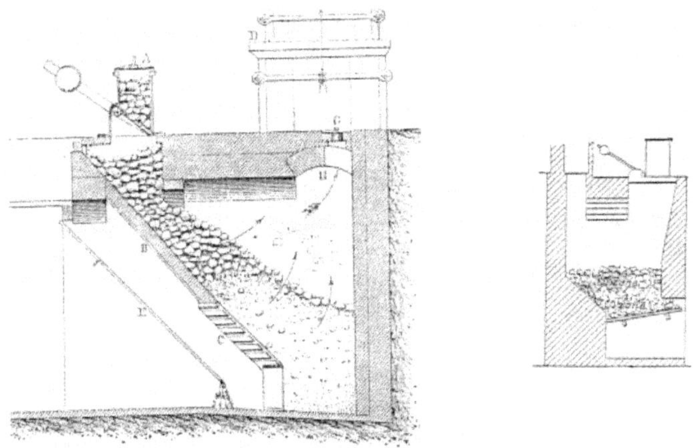

Figure 18. ***The Inclined Grate Gas Producer***

The drawing taken from Henrivaux shows a deep bed of coal on an inclined grate. Coal is fed through the hopper and valve on top of the producer. Water is trickled into the hot ash bed where it evaporates and is comverted to gas by the coal fire. Air is blown under pressure through the grate and the gas fed into the furnace via the offtake gas flue

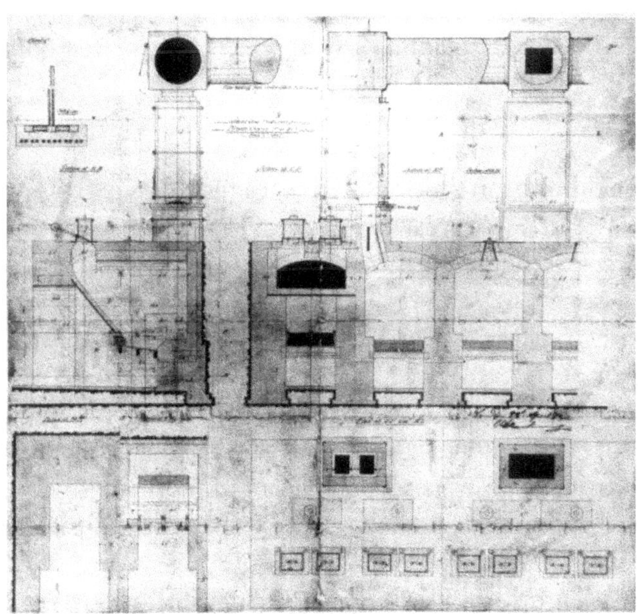

Figure 19. ***This type of producer was in use at Chance Brothers as shown by the photograph of an old drawing***

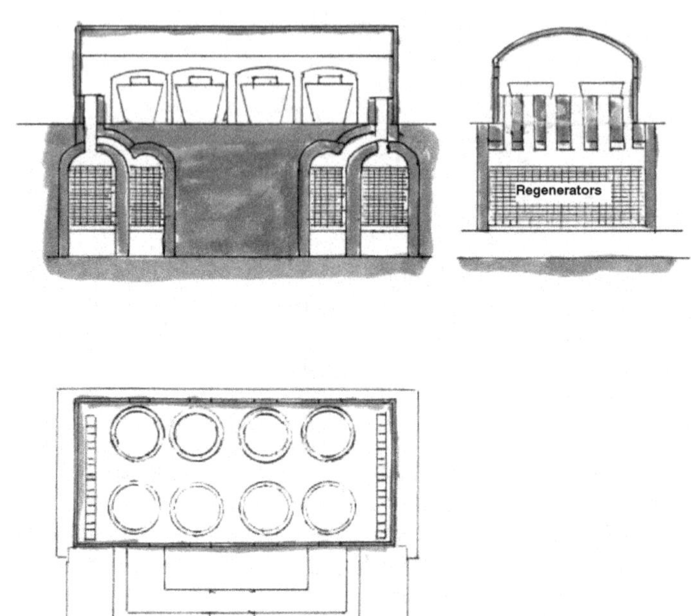

Figure 20. ***A Diagrammatic View of an End Fired Gas 8 Pot Glass Melting Furnace***

By 1863 Pilkington's adopted the regenerative system for their Rectangular Pot Furnaces. This meant that the furnaces were re-designed to burn gas, and the indications were that the company now had a separate gas producer. In 1864 the application of steam improved the performance of the producer and in the company they were pronounced a success along with the gas firing of the kilns and auxiliary plant. There was some trouble reported in November 1863 with the reversing valves, and no wonder since they were simple butterfly valves made in wrought iron. There were considerable savings and the pots wore better. The first gas fired rectangular pot furnace could have been centre fired, but later were most likely to be end fired as reported in August 1867 when they altered No. 6 by putting the flues at the ends. In an end fired furnace there were four parallel stacks at each end and the gas and the air mixed inside the

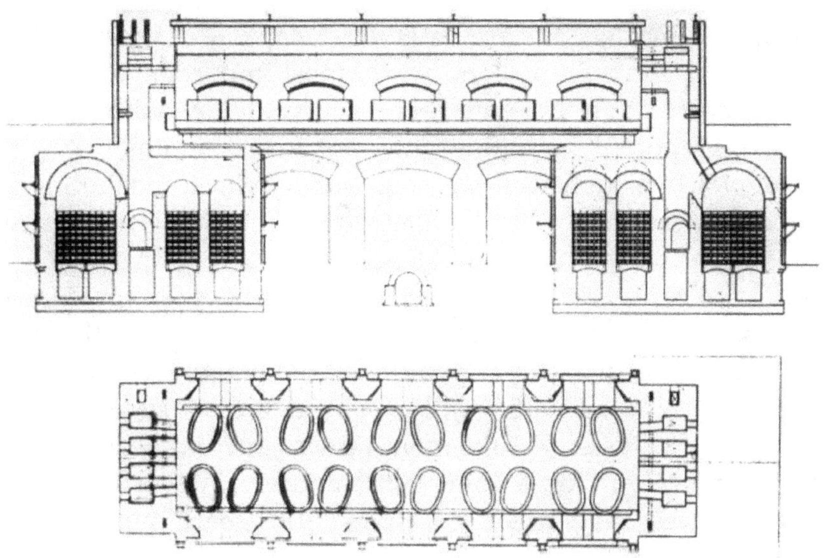

Figure 21. ***The Bicheroux Pot Furnace — Plate Works***
This furnace was operated at Plate Works right up to relatively modern times. It was an end fired gas furnace exactly on the same lines as that used in Sheet Works for melting and making sheet glass in pots between 1863 and 1873. The pots were loaded on to the sieges from each side through openings under a relieving arch, and closed by large tweels. They were filled through an opening in the tweel. The pots were very similar to those used in the past but were oval in shape. They were lifted in and out of the furnace by a fork that fitted into a slot incorporated in the side of the pots very similar to that used for lifting the 72" pots in 1870

furnace and burned with a lazy flame. The construction and working of the Bicheroux end fired furnace at Plate Works in the 1950s supports this contention that end firing was the most likely method to be adopted, Figs 20 & 21.

This not only proved to be a success in the firing of the furnaces but enabled the company to apply gas firing to the kilns and the pot arches. One immediate effect of gas firing a rectangular pot furnace was that clay crowns were now exposed to much higher temperatures causing them to melt on the inside and give rise to drops and strings.

In 1863 a preference was expressed for a gas furnace in No. 1 House with a Dinas crown and Yorkshire bottom, which was completed and lit in April. Dinas was a silica based brick capable of withstanding much higher temperatures and was a most important advance in furnace construction. In September 1863 No. 5 Yorkshire firebrick crown was reported "dropping" after only three months.

In July 1863 "thought well of gas furnace, four in all and putting up two more. Saved 25% on coal" and in November 1863, "gas furnaces decidedly the best, continue to do well".

At this point in time, in May 1864, the gas furnace on No. 1 House was apparently making "wretched" metal and consideration was given to replacing it with a 12 pot furnace. Although it continued to operate up to November 1864 with varying degrees of success, it was reported that at the end of the month it was out and nearly ready. In December No. 1 House doing better and "W. W. Pilkington still hopes of success on her new principles." The Board minutes from December 1864 to June 1865 report that after starting "Lighted No. 2 side but not made glass in her yet. Made no glass in the other side since Monday last, the division in the furnace now removed." "Doing fairly, only one side going yet expect to light the other on Monday next." " One side doing well and so far a success. Hope to light the other side early next week." "Working both sides and metal moderate. Rather inclined to seed". "Furnace still slow but quality better than from others". When it came to March 1865 W. W. Pilkington spoke at some length upon the merits of No. 1 as compared with the other houses giving it as his opinion that "she did not answer as well as she did when she was a 12 pot furnace and that he thinks it would be better to turn her into a 10 pot and her old form rather than continue her as at present or incur further expense with her." It is only possible to speculate what this was all about. The possibility was that it was some alternative form of gas firing, but it was only in 1872 that there was a drawing of a centre fired 10 pot furnace prepared for Messrs Pilkington, the pots being loaded in at the ends. As whatever it was, did not work satisfactorily further speculation might be pointless, but it could be said that a centre fired furnace could suffer from short circuiting in the centre whereas an end fired furnace developed a flame over its full length. Anyway in May 1865 there was a report that No. 1 had gone wrong with something pos-

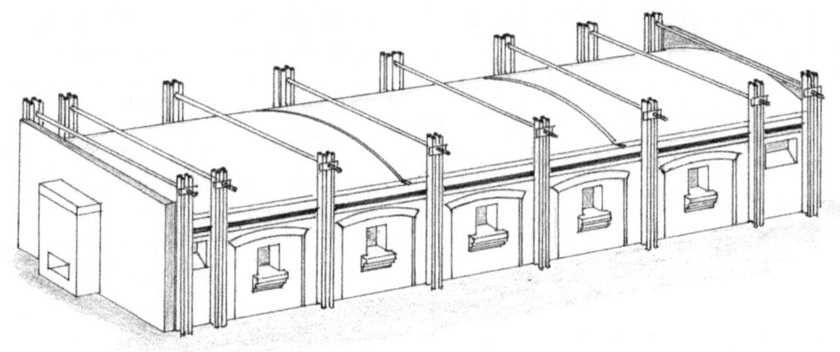

Figure 22. ***A 10 Pot Gas Fired Glass Melting Furnace***
The drawing shows a gas flue on the end for heating up the furnace from cold to gassing temperature as well as the pipe holes for warming the pipes. Scale approx. 1:60

sibly connected with the flues so that in June it was decided to put her out and alter her to an ordinary 10 pot furnace, Figs 22 & 23. However, this was amended a week or so later to an 8 pot because her three producers would be inadequate if worked with slack, which was cheaper but less efficient.

There was no further information about the furnaces for the whole of 1866, 1867 and 1868, except for a note that in July 1867 No. 1 was started up with 12 pots. In November 1868 Siemens advised against a 14 pot furnace preferring in principle, a 2 × 8 pot furnace. However, in February 1869 there was a report "Many of them giving way in the crowns. William Pilkington and Hartley thinks that the shape has been at fault and that the bricks have not been sufficiently keyed in the centre of the crown". This may of course have been due to the expansion of the silica in the crowns and more pronounced as the pot furnaces got larger.

In 1869 there was a succession of reports about first and second halves of No. 6 and No. 2 Houses, for example, 2a and 2b as well as 6a and 6b. These were referred to as double or coupled furnaces. It is reasonable to assume that during the three years from 1866 to the end of 1868 there were developments that were not recorded. It was a fact that a 12 pot furnace was nearly at its working limit. There was

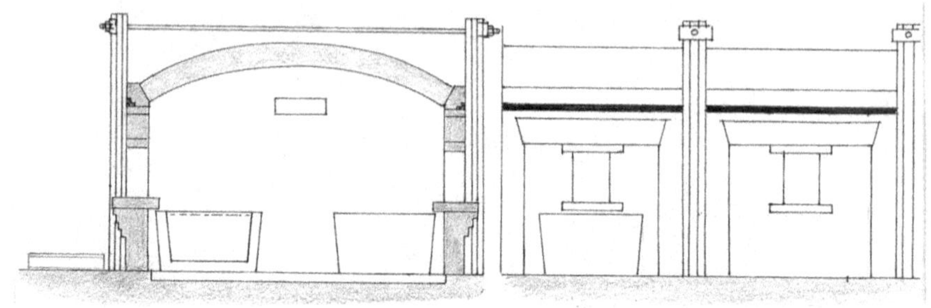

Figure 23. ***A section through a Gas Fired Pot Glass Melter showing the teazing and gathering holes using a pat height of 2' 3"***
Scale approx 1:30

mention of a 14 pot furnace, but the two pots at the ends were small. A 16 pot furnace would have been out of the question, for it would have been beyond the limits of design and practicability. The question was how were the furnaces coupled, was it by a common producer, a chimney, or regenerators? It was Dinah Stobbs who suggested that it could be alternate foundings that coupled the furnaces and this suggestion supplied the complete answer. Two 8 pot furnaces coupled together would double the number of foundings and make much more glass in a week than a single 16 pot furnace. The design of the double furnace could be to have two single furnaces end to end with their own sets of regenerators their own sets of producers, and one chimney. Foundings would be alternated from one furnace to the other using only 16 blowing holes and not 32. In 1870 the first reports of dust as well as sulphur appeared. This continued through 1871 and was finally related to the wheeling of frit into the Houses whilst the glass was being worked, thus confirming the nature and the working of a double furnace. Thus it can be said that in the period 1869–71 the productive capacity of the company was doubled, and a landmark in its history.

From 1871 the time was occupied in improving the speed of the gas furnaces. Apparently they were not as fast as the coal fired ones. By speed they meant that the power of a furnace was less and that it took longer to found the pots on a gas furnace than on a coal fired one. In June 1871, the first gas furnace in No. 6 was faster and safer than any subsequent ones because the gas came in at the teazer hole ends with long slits and high fender. It is difficult to say whether this inferred that it was centre or of an end fired design. Modifications were made in September1871 to No. 1 and later No. 6c with a much larger gas inlet and modifications to the regenerators. This produced a good solid flame and was safe provided care was taken. All told this reduced the founding times from 29–30 hours to 23–24 hours. There was and always has been a good chance of explosions when reversing a regenerative furnace, so that by safer they mean that, with care, there was less chance of blow backs that could lead to breaking the pots.

At first a central bank of producers supplied gas to the tanks as well as the kilns and blowing furnaces. In June 1871 "consider it expedient to put all the producers behind No. 2 house. For many reasons it would be an advantage to limit the number of kilns to a certain number of producers, instead of all connected to all." Thus it was that in future each tank and each bank of kilns had their own producer.

CHAPTER 5.

Pots and Gathering Rings 1826 to 1873

The claymaker and claymaking was an art and a skill inherited from ages past. It was also an inherent part of glass making. Apart from the making of the clay bricks used for building the furnaces, regenerators and flues there was the specialist claymaker who made the pots, rings, floaters and kiln claywork, Fig. 24.

Pots were made from the best Stourbridge clay. The clay was, in later years, weathered on a 2 acre, brick paved field, at the Mill Lane siding at the Rainford sandwash. It was pugged in the works but there has been no indication when grog was used or when ball or other mills were available to make the grog. It took about 5 days to make a pot, and when completed it was left to dry for about a year at 60°F increased to 90°F until all the water had been driven off. They were then fired in a pot kiln or pot arch. In the pot glass melting furnace enough cullet was melted in the pot for the teazer to glaze it on the inside before the pot was filled with frit (see Richard Pilkington).

Rings were made in much the same way. They were always put in the bottom of a pot and allowed to float to the surface at each successive found.

The life of a pot, and for that matter, the ring, was about 8 weeks, and the cost of a pot was £9.

Prior to 1826 and up to 1834

It is thought that the early pots used for spread and crown glass making held about 450 lb of glass. The round pot furnace installed in the Old Cone or No. 1 House used pots of about 600 lb capacity with 16″ diameter rings.

1834 to 1845

The early rectangular pot furnaces making crown glass used a 32″ pot which held about 1000 lb of glass. One pot would make 100 tables and each gather weighed 9 lb so that it would need pots of this capacity to do this.

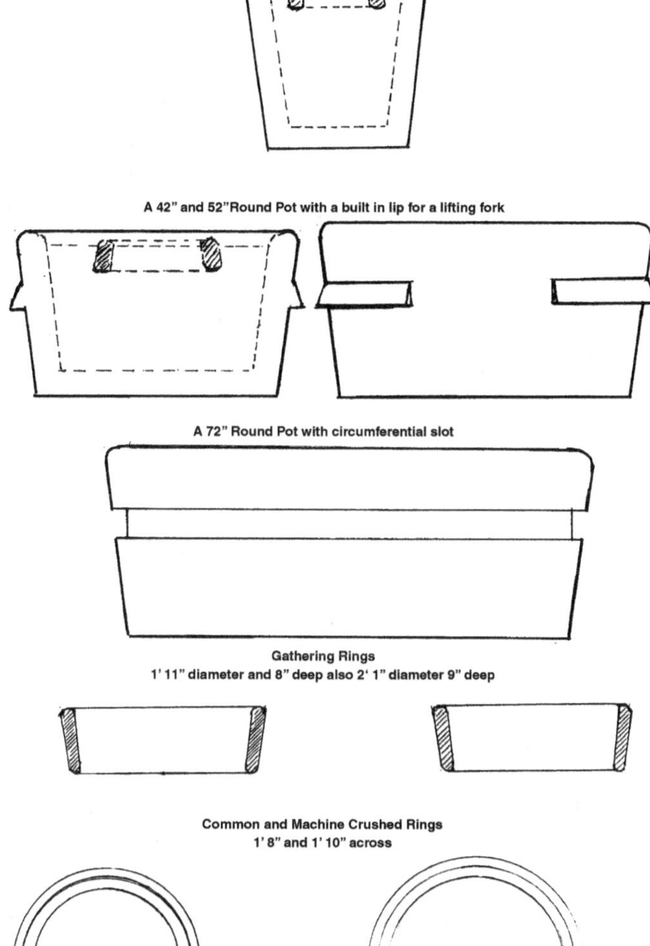

Figure 24. ***A selection of pots and gathering rings***
26" Round Pot with 16" Gathering Ring
A 42" and 52"Round Pot with a built in lip for a lifting fork
A 72" Round Pot with circumferential slot
Gathering Rings
1' 11" diameter and 8" deep also 2' 1" diameter 9" deep
Common and Machine Crushed Rings 1' 8" and 1' 10" across

1845 to about 1870

Crown glass was being phased out and replaced by hand blown sheet. The first pots to be used were the 42″ ones of about 1500 lb to 1800 lb capacity called the small pot. In 1860 the coal fired pot furnaces were replaced by gas fired pot furnaces and improvements in coal firing as well as gas firing enabled the small pot to be enlarged, and the 52″ pot with a capacity of 30 cwt adopted. These large pots used rings of 18″ diameter 2″ thick.

The early pots of up to 32″ diameter were handled by sliding them in and out on their bottoms. The 42″ and 52″ pots were too heavy so a half round lip was incorporated on the sides so that they could be handled by a fork on a two wheeled truck (Ref. Tank Drawing Books, Pilk. Archives PB95).

1870 to 1873

Towards the end of the life of gas fired pot melting furnaces and before they were replaced by tanks, the 72″ pot was introduced. This pot held 2½ tons of glass but it was so heavy that a circumferential slot had to be incorporated in the body of the pot so that it could be handled by a half round fork on a wheeled truck

A study of the pots shows that the increase in capacity was not so much gained by increasing the depth but by increasing the diameter. Small increases in diameter had greater effect in increasing capacity than increases in depth. Not only that but increases in depth would have made it more difficult for the gatherer. The influence of the gatherer would therefore appear to have influenced the pot design and given rise to pots of nearly constant depth with ever increasing diameters.

Gathering Rings

Eighteen inch and sixteen inch gathering rings were used in the pots. The same size of rings would most likely have been used in the early tanks and up to the first divided tanks in 1883. Later in 1893–97 when the large divided tanks came into use the size of the rings were then about 2′ 1″ in diameter and 9″ deep. The crushed rings had a flat incorporated on them no doubt to give them stability when located under the gathering holes, against the side blocks. These rings were also used in the melt end along the division wall to form a barrier to protect the side blocks against erosion from the flux (Ref. Tank Drawing Books. Pilk. Archives. PB95).

CHAPTER 6.

The Transition from Pots to Tanks 1870 to 1880

The Large Pots

Large pots were tried from the end of 1871 to the beginning of 1872. Early opinion was that the large size did not lead to good quality. No. 2B gas furnace was purposely built in March 1872 for 8 large pots but did not do as well as was expected. The quality was not first class or as good as it would have been in ordinary sized pots, but the quantity produced was equal to the ordinary production of almost 12 ordinary sized pots. In view of this, it was "decided to try small pots first journey in A House when she is lighted up and to use up to any of the 36 pots now ready." "It will also test the time of founding such sized pots and other such points of use whenever we do work over pots." Later on in April 1872 , "2B was doing well with all large pots and getting factory metal out of them all". Unfortunately, this success was more related to an improvement in the better sand and better washed sand. Up to the end of 1873 when the first trials of the continuous tanks started, the arguments for or against large pots revolved around the number of founds. It soon became apparent that it took longer to found a large pot than a small one and it finally transpired that for quality and production it was doubtful if there was much to be gained from the use of large pots. According to a report from another glass making factory a large pot there was 73" in diameter. This pot was truly enormous and must have been very difficult to handle.

Large pots were still in use in 1874 so it can be concluded that they were not entirely a failure. In one case at No. 6 where they were not doing well in February 1874 W. W. Pilkington blamed the management of that house.

The Day Tank Furnace

A Day Tank (Fig. 25) was one that would be filled and emptied over a period of one found in the same way as a pot, and like the pot melting furnaces the actual production of glass would be limited to 3–4

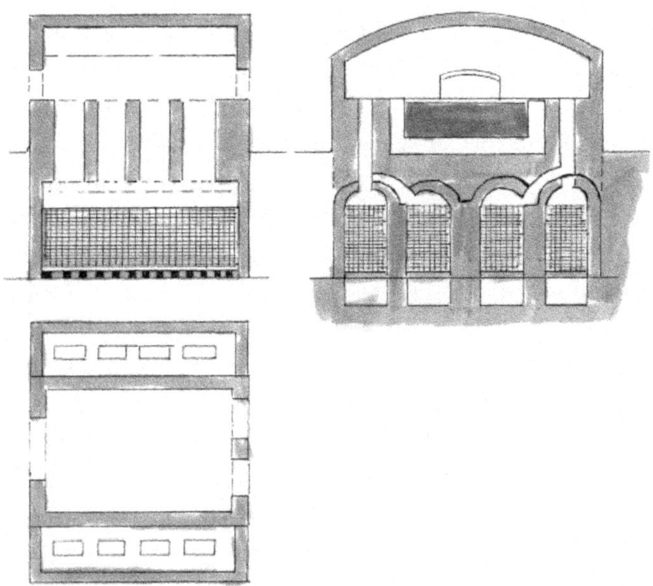

Figure 25. ***A gas fired day tank***

This is a diagrammatic drawing of a cistern or day tank. The cistern replaced all the pots and it was about 10' square. It was gas fired from the sides still using the lazy flame. The cistern was built on top of the regenerators in the same way as a pot furnace. It was filled, the glass melted and emptied as if it were a pot. There were indications that this type of day tank was experimented with by the Company but there could have been problems with the bottom blocks overheating

founds in each week. Even though there were limitations, there was always some incentive to melt glass in a tank instead of a number of pots and pots were getting so large that it would appear to be only a small step to a tank. Pots were subject to breakage and wear or thinning as it was called and there was some advantage in having one tank instead of 8 to 10 pots. This was pursued more on the continent in Dresden than in Britain, and it was possibly the first step towards a continuous melter. In March 1863 Siemens was said to have taken out a permit for a furnace without pots. It would appear that Siemens work on day tanks was directed more to the bottle makers and there was an indication that day tanks were used by bottle makers (see Siemens patents).

In Pilkington's there was talk of a cistern (or day) tank in June 1868 in No. 3 House. But in October 1868, partly owing to a shortage of glass, the idea was abandoned and the bricks for the furnace taken off their hands. Despite this in June 1869 it was said that the blowing furnace worked much better than "our blowing and melting furnace all one". It has not been possible to interpret this remark in the records but in October 1869, "debated plan for putting up a second tank and kiln next to present plate house at a total cost, enlarging the buildings, 5 kilns in all £1000". Such a tank furnace would probably hold 181 cubic feet of glass, whereas an 8 pot furnace would hold 204 cubic feet and a 6 pot 198 cubic feet of glass. The size of this tank would therefore be about 10′ square. There was no confirmation that it had been built, but in June 1871, "tank has been bad but last week improved". The conclusion was that although there was little mention of the day tank between 1863 and 1872 there were indications that something had been done, not only on a cistern tank, but a cistern tank used for gathering and blowing. Although only two tanks were specifically mentioned there could have been a build up of experience and information on the working of day tanks before the next step was taken to build a continuous tank in 1872–73.

The Transition

Up to the period between 1863 and 1873, glass had been melted in pots. Melting glass in such a receptacle as a tank or cistern would have meant the construction of a very large pot. A pot of that size would be impossible to burn, so pre-burnt blocks of clay would have to be used. As the company was expert in making pots it is reasonable to assume that the tank would be built like a pot, so that when the blocks were assembled they would have been grouted in with clay along the bottom and up the sides. Going back to the beginning in 1863 the Company was at that time buying in alumina furnace bricks and siege blocks to build pot furnaces, and clays to make pots at Eccleston Street. As the siege blocks were of about the right size they could have been used to build a cistern tank. In 1873, the Siemens bottle glass tanks were built of large bricks made and baked in the glass works. They were joined by a special mixture "containing very little silica" and were nearly white and rather soft (somewhat like our K32) when baked, but after much use in the furnace they became red, very scaly, and much more friable. It is also curious that in July 1874, after the first continuous tank was built in August 1873, "No. 3 tank so full of dirt that she is being ladled. Think the clay packing between the bottom and the sides has been coming up". Earlier in April 1873 the first tank had failed after a good start and according to a report in Barker's book this failure was due to the bottom having been eaten through after only one week. Between Sept. and November 1873, the second tank ran well for 97 days. In May 1874 3b tank after 2 months of working, "gave way in the bottom and had to be ladled", then in July 1874 the packing came up. With all this evidence it would support the idea that the cistern tank was made of blocks and parged with clay and that this method of building tanks continued to be used for the new open tanks later on. It might explain why it took so long to develop a day tank and even if one was developed it was not reported to be a success.

In 1875, Siemens reported that arrangements for ventilating the bottom and the sides are such that a red heat was not observable in the ventilating flue and in his newest designs he had much improved the ventilation of the tank and provided for any runs of molten glass that might accidentally occur, to fall into the open cave and not in the narrow flues, Fig. 26.

It is a fundamental fact that the continuous melting and making of glass in a tank is dependent on the molten glass freezing between the joints of the blocks before leakage occurs, and this is why steel making is not a continuous process.

Reading between the lines it was reasonable to assume that Siemens in Dresden between 1863 and 1873 must have realised something about the characteristics of glass freezing between the joints in a tank made of clay blocks. However, it was possible that the Company did not recognise it for some time and this might explain why they were so slow in this period in developing the cistern tank and why they continued to think in terms of pots when they built the first tanks. It was not clear how the bottoms were laid or constructed in the early tanks but it was known that the sides were very enclosed by the stacks and buttress walls, the crown being built on and sprung from the blocks but what evidence there was would point to the possibility that the early tank sides and

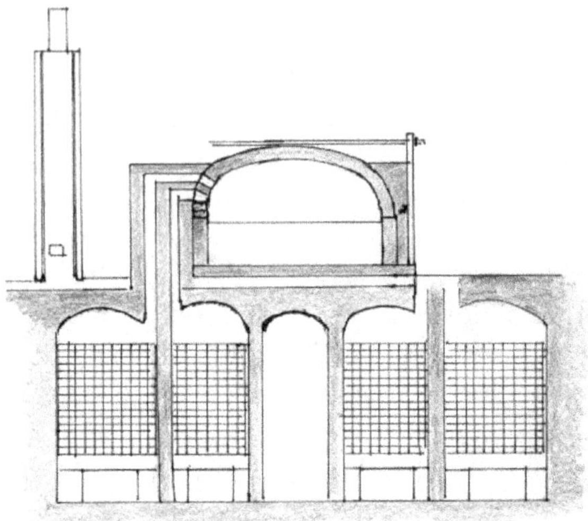

Figure 26. ***A cross section through the tank showing the regenerators, the stacks and the induced draught chimney for cooling the bottom blocks***
Scale approx. 1:60

bottoms were lined with clay, in other words, in concept, built like a pot. It is no wonder that there were reports of tank failures, heavy wear and clay coming up. After experience with the first tanks and after July 1874 there must have been a radical redesign, not only to accommodate crown expansion by springing the crown from a springer plate, but by introducing better bottom cooling and removing the clay lining.

Tank Blocks

This was an investigation to confirm what blocks may have been used to build tanks, when tanks first came into use in 1863 as small cuves or cisterns, and later in 1873 as small open tanks, and then to follow the investigation up by recording the history of block making up to 1952 as recorded in Reminiscences in the Pilkington Archives.

In the period 1890–97 the need for pots was nearly over and in 1891, a decision was made to move the pot making premises from Eccleston Street to Sheet Works. Up to that time no records had apparently been kept and any formal records only started to be kept when they went to Sheet Works. Charley Woods thought that we bought all our blocks before 1897. In Herman's (the works chemist) time there was a record of a talk with suppliers that states that none of their blocks were as solid as those produced by Pilkington Brothers. Herman and Geoff Hyde in ... developed and made our own tank blocks "fed up" with being let down so often. Frank White the claysheds manager avers that few tank blocks appear to have been made until 1902 most of them having been bought from Stourbridge and many from George King Harrison. In 1876 a tank furnace at Pilkington's the body of which, was reported to be built of firestone from North Wales and cold air drawn around the hottest part of the tank to prevent the metal from coming through.

The explosion in the development and building of open tanks occurred in 1873 in conjunction with Siemens and it does seem to confirm that the company went to the brick makers for their blocks at that time and not have them made at Eccleston Street.

In 1904 Sheet Works made moulded blocks using the K32 clay mixture and for the record:-

Tank block mixtures were designated	"K" e.g. K32
Tweel blocks	"U"
Siege clays	"E"
Clay rings and floaters	"R" e.g. R43
Mending clay	"M"
Stoppers	"S"

In the larger tanks for making machine drawn cylinder glass the K32 blocks were not good enough for it must be remembered that the depths of these tanks were now 4–5 feet and not 3 feet as in the blown tanks. Thus it was that in 1930–32 a new process for making tank blocks by machine was adopted. These were designated "S&G" blocks. They were made from a hammered damp powder mixture of ground clay and grog in steel moulds using pneumatic hammers.

S&G was superseded in 1938 by an Al.17c slip cast block. Although slip casting was experimented with in 1916–18 it was not brought into full use until 1940. This block stood us in good stead throughout the 1939–44 War with Corhart and Sillimanite coming gradually into use. After 1946 saw the greater use of Corhart and increasing use thereafter of electrocast refractories.

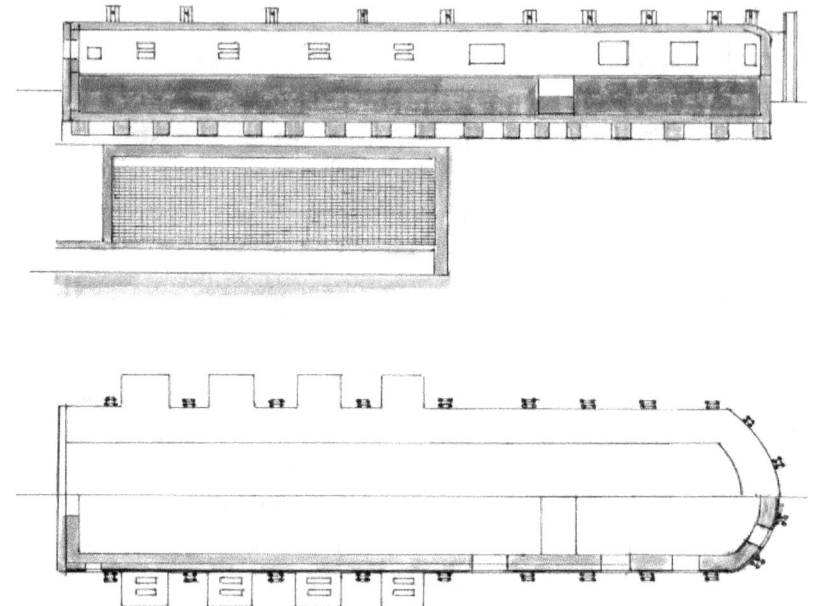

Figure 27. *The first Continuous Open Tank with two gathering holes, four blowing holes, and either a fixed barrier with a doghole or a floating barrier across the lower end of the Tank, as deduced from various sketches and descriptions*
Scale approx. 1:60

The Continuous Open Melting and Making Tank

Up to 1872 all the glass had been made in the Rectangular Pot Melters apart from any that may have been made from a cistern or day tank. In July of that year a continuous tank was proposed on the principles of the one that had been seen working in Dresden, Figs 27 & 28. This type of tank was one which melted, fined, and prepared the metal for working in one continuous operation. It meant that glass could be produced continuously throughout the week; a very big step forward.

In August 1872 Siemens was in London to discuss suggestions as to the rings and divisions in the furnace. In April 1873 the tank was put into production for the first time. The glass was reported to be slightly seedy, but it continued to improve and was beating any pot furnace on the ground, having very few blisters and defects of that

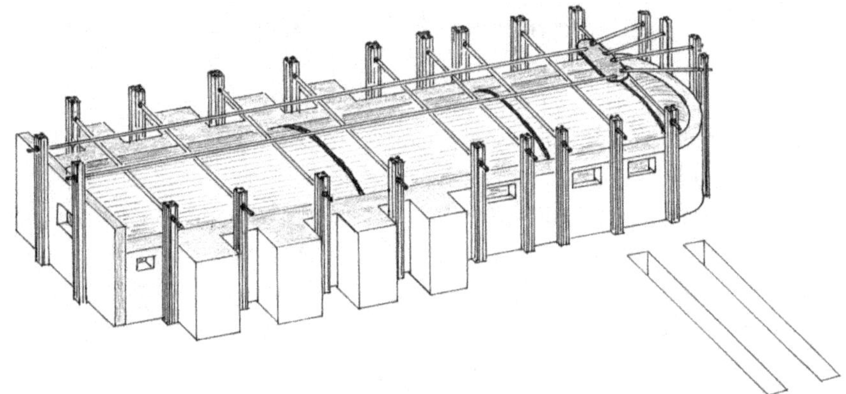

Figure 28. *The first Continuous Open Tank shown in perspective with the swing holes*
Scale approx. 1:60

kind. It was so good that it was proposed to discuss the advisability of getting the patent secured by Siemens for sole use by the Company.

There were no further reports for some time but in August 1873 it was reported that the tank was lighted and making glass for the second time and "doing fairly". There is an interesting note in Barker's book to the effect that the first tank failed soon after the very good start and that it was a constructional failure. This may not be very surprising considering their limited experience in the construction of tanks and the problems that had been experienced in the transition from pots to tanks.

The second tank ran well for 97 days. There was some wear from the flux on the pipe bridges but the bottom was only worn 3" in depth. The projections into the furnace were also worn. What was meant by the "projections" was is uncertain. There was a later report that the "crown bricks at the gas holes has rather been against her," i.e. possible wear on the ports. Decided not to renew the bottom but to build the tank with one bridge instead of two. There was apparently some wear on the fluxline at the bridge, no doubt because the flux was allowed to get too far down the tank. The solution proposed was to put rings in on the melt side of the bridge as well as the rings under the gathering holes. Later it was reported that the new tanks were working well and that No. 4 had lasted 7 months and No. 4a, 8 months.

The first continuous tanks according to Barker, were short and wide, but very soon after were reported to be built, longer and narrower. The reason for this may be that in a short tank, it would have been very difficult to grade the temperature required at the melt end down to the lower temperature for fining and working the glass at the working end. In addition a longer and narrower tank would allow the frit to be progressively melted as it went down the tank, so that by the time it arrived at the bridge melting would be complete. Circulation of molten glass may not have been a factor in the thinking in those early designs. The fact that these tanks worked at all was probably due to the low draw. The tank in the fifth House once had 4 feet of metal. It was known as the Belgium tank but it proved a failure because it could not be got hot enough and had to be done away with.

It was interesting to note that the gathering end was rounded, despite the fact that they had for some years now gathering had been from a straight sided pot furnace. Most likely this would be to give the gatherers more floor space, for gatherers needed more room to work sideways than blowers.

After an eventually successful start Pot Melters were rapidly replaced by tanks. The next tank in No. 3 House was a larger one followed by a tank in No. 5 house in May 1874. There was a curious reference in February 1874 to a tank in old No. 5 House being now out. Could this have been a cistern tank?

In July 1874 there are references "Dinas bricks expanding more than Stourbridge which contracts". There were also references to springer plates and iron bracings. This was interesting for it meant that tank crown expansion was a serious matter and had to be taken into account. The first tanks had two gathering holes and four blowing holes and they were built in the same way as the pot furnaces where the crown was sprung off the siege side walls they were now sprung off the side blocks, Fig. 29. Although bottom cooling had gone some way towards solving any problems in that area, this form of construction, and the enclosure of the side blocks, could have lead

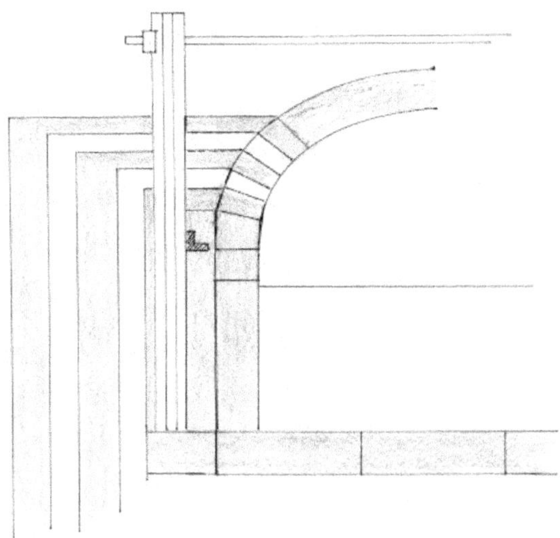

Figure 29. ***A cross section showing in detail the stack, ports, and the spring from the side blocks***
Scale approx. 1:30

to potential failure. Some little time later the tanks were enlarged to accommodate four gathering and eight blowing holes, and by 1877 at least 12 tanks had been built. It was possible that by about that time, these larger tanks were built by springing the crown off a springer block on the springer plateand moving the stacks away from the side blocks. There was no indication in the records whether this important change in design took place then or after the tanks were divided in 1880–81, for these new divided tanks did have a springer block sprung crown at the melt end, even though the working end crown was still sprung from the side blocks.

The Operation of a Continuous Tank

In a continuous tank, glass was gathered, and then handblown over the metal in the gathering end of the tank. At first there were two gatherers and two hand blowers, but by 1880 the tanks had been enlarged to accommodate four gatherers and eight blowers.

The operation of a continuous tank was quite different to that of a pot melter. The continuous melting and making tank meant that the mix of batch and cullet could be filled and melted continuously to keep the gatherers and blowers working on presumably three shifts all week, shutting down only at weekends for founding, repairs and burn-outs.

The frit and cullet was wheeled into the tank house in barrows and filled into the tank through a hole at the melt end. Wind storms caused dirt which arose from the quantity of frit which trickled down from the barrows, a fact that confirmed this new way of working and making glass. In October 1875 it was reported that "men getting used to tanks. They were thrown out of kilter when they first went from pot to tanks but are now getting better at it". From 1874 through 1875, and up to 1878 it was a struggle to maintain quality in nearly all of the tanks and it was only in March 1878 that there was a report that on the whole "tanks doing well and pots badly". It was just before this that it was reported that the glare from the tanks was affecting the eyes of the workmen and it was recommended that blue or green spectacles be used. Finally in January 1879 it was decided to break down the stock of pots by degrees being quite satisfied to rely on tanks entirely.

In January 1877 we have the first references to tank managers, for the tanks built between 1873 and 1880 were open tanks and they would be very difficult to operate. Like some of the Siemens patents the difficulty would be in balancing the temperatures to get melting at one end and working temperatures at the other to say nothing of fining. The low draw on the tank may have helped.

Sulphur was an ever attendant problem although much freer than it was in the old pot days.

CHAPTER 7.

The Making of Hand Blown, and Compressed Air Blown, Sheet Glass from 1850 to 1880

(ref. Henry Deacon's paper Feb. 1851)

Hand Blown Sheet Glass

Hand blown glass known as sheet glass was a big improvement on crown glass for quality and thickness. The improvements were in the size of the cylinder, and the fact that 40″ × 30″ panes of glass could be produced with much less waste. The improvement over broad glass was that the cylinders could be cut cleanly with a diamond and not sheared.

Hand blown sheet glass was made in varying thicknesses and measured by weight. The thinnest glass weighed 13 oz per sq.ft, the next 16 oz, 21 oz, 26 oz, and 32 oz. In millimetres this was equivalent in thickness to 2 mm, 2.3 mm, 3 mm and 3.6 mm.

Using a lighter blowpipe than for crown the gatherer, in much the same way gathered, 8 or 9 lb or up to 30 or 35 lb weight of glass depending on the thickness of glass to be made. He then formed the metal on the end of the pipe and shaped it in a hollow block of wood called the deep mould block, Fig. 30.

The blower then took over and by blowing shaped the gather on the rounded ready for blowing into a cylinder. The blower then went to the blowing furnace which had taken the place of the flashing furnace. There the blower heats the piece and by blowing and swinging it in the swing pit, it starts to elongate. After a second and third reheat a cylinder with an even thickness throughout was formed. Finally the end of the cylinder was heated whilst the air inside was trapped. The trapped air expanded and blew out the end of the cylinder. The side-walls at the end fell in and were reheated so that when the cylinder was spun on its axis they opened out and were then fire finished. The method used for the thick cylinders was to dab on to the end a blob of metal. Again by trapping the air inside the blob of metal was blown off and the ragged edges cut off by shears and fire finished.

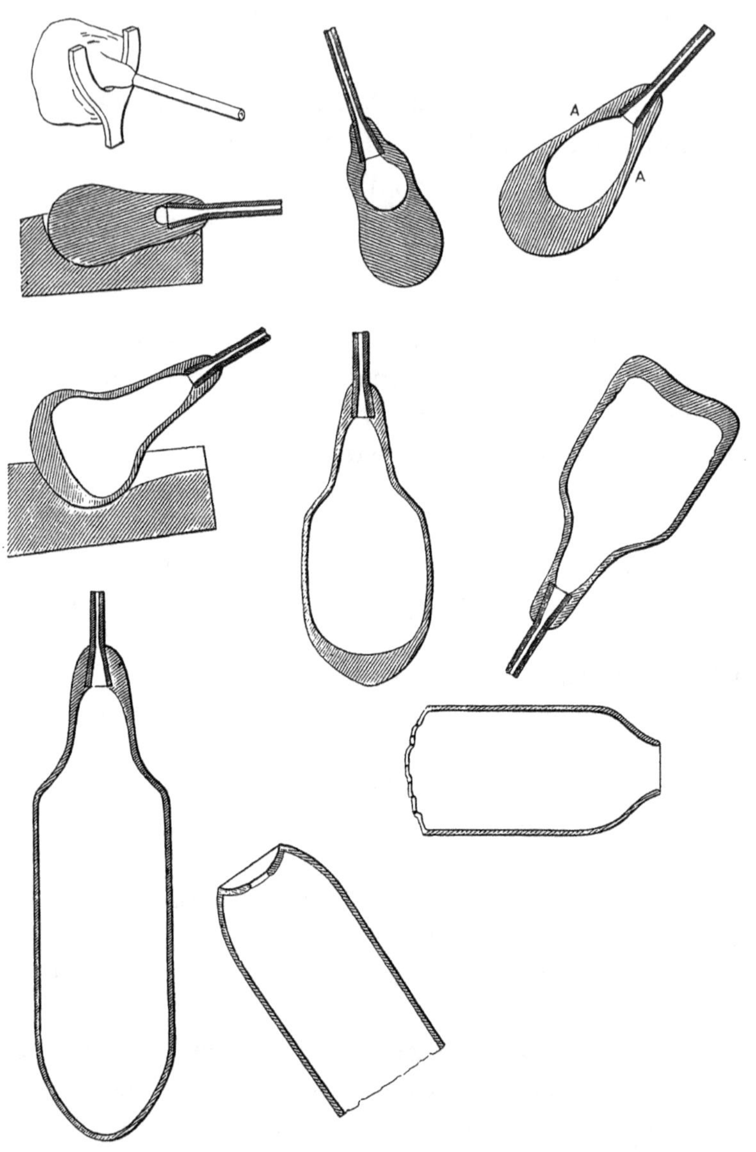

Figure 30. ***These drawings, taken from Henry Deacon's paper on the manufacture of blown window glass show the sequence of operations from the gather to the shaping and finally the blowing of a cylinder***

The cylinder was placed on the chevalier and the blowpipe cracked off. Finally, a strand of hot metal from the cordeline was wound around cylinder and the cap cracked off leaving a cylinder open at both ends and of even thickness throughout.

The men work from 8 to 10 hours each time with ½ hours rest in the middle of each journey. The team can make from 90 to 115 thin cylinders and proportionately less for the thicker ones.

Compressed Air Blown Sheet Glass

(as described by Harry Langtree TD in his lifetime)

The making of sheet glass by the hand blown process lasted until compressed air was introduced in about 1870. At about this time in 1869 iron forming blocks promised well using a starch and charcoal or potato and charcoal mixtures. They were formally adopted in May 1870.

The use of compressed air enabled even longer and larger cylinders to be made as well as a greater variety of thicknesses. There were at about the same time improvements made to help the blower to cope with the extra weight of the gathers in the form of what was called the bicycle carriage. This was a frame on wheels which took the weight of the blowpipe and the gather. It allowed the pipe to be swung in the swing pit and the cylinder to be heated in the furnace with much less effort.

The following is a description of the production blowing process as carried out on a continuous blowing glass furnace, from 1886 onwards and there was little or no change from the procedures carried out long before this date or for that matter long after.

At the working end of the glass furnace there were up to ten blowing and five gathering holes. At each gathering hole there was a gatherer, a time gatherer, and a block minder, Fig. 31. The blockminder was responsible for preparing and maintaining the blocks, the tools, servicing the gatherer, and helping with the compressed air. The tools of the trade at the gathering end were blow pipes about 6′ in length, and the cast iron blocks. The pipes could be used for hand blowing or with compressed air. These pipes cleaned of scale by the block minder and warmed in the pipehole ready for gathering. The time-gatherer

Figure 31. ***Gatherer's gathering the weight***

made two or three gathers before handing on to the gatherer. The gatherer was responsible for the final weight and the first forming on the deep mould block. The deep mould block, Fig. 32, was used to form the gather and the rounded block to form and set the diameter at the shoulder, which in turn, set the diameter of the cylinder for the blower. The blocks were lubricated with graphite or a sprinkling of sawdust which when burnt off left a layer of carbon. The rounded blocks were water cooled between gathers and there was a stream of water for cooling the pipes as the heat ran up them. Even though water was used the gatherers had to use a piece of hand held curved metal backed with leather for holding the pipe. When not in use it was attached to their belt as was a towel to dry the pipe. The gatherers used the same clothing as the blowers except that, the blowers had their shirt tails hanging out, whereas the gatherers had theirs tucked in under their belt. They all wore light check clothing supplied by Tyrers of Bridge St. and this clothing was taken home in a bundle after work.

When the gather had been completed the pipe was handed over to the blower. The blower's basic tools were, a pick, shears, pinchers, cordeline and callipers. His job was to heat and blow the cylinder which he did in the swing pit, after a succession of reheats in the furnace. It took three reheats in the furnace, Fig. 33, before the cylinder

Figure 32. ***The blower shaping the gather***

could be blown to its full length. It was important that the pressure in the cylinder was maintained at all times. The fourth heat was on the end of the cylinder to enable a ball of hot metal on the cordeline to be dabbed on to the rounded bottom. After the fifth heat the compressed air was applied so that the ball and the surrounding metal could be blown off into the tank. Any surplus metal was then sheared off by the block minder and after another quick heat the cylinder was spun to open out what was the bull end. Finally, the cylinder was laid on the chevalier and the blowpipe cracked off for it to be cleaned and returned by the block minder. A string of hot metal gathered on the cordeline was wound round the cap end and using the pick the cap was broken off. The cylinder was marked in common chalk with the blowers number on the inside whilst on the outside it was marked with the same number in French chalk which burnt in during flattening. The finished cylinder was taken from the chevalier by the carriers and stacked in the "palace" ready to be taken to the splitters for

Figure 33. ***The blower heating the cylinder on the bicycle machine***

examination and splitting. Palace was the word used for "the place where" cylinders were stacked, ready for taking away on day turn. Glass was blown at various weights, 9 oz, 11 oz, 15 oz, 21 oz, 26 oz, 36 oz and 42 oz, solely by the skill of the gatherer and blower. The record for the production of photo glass was 201 cylinders in 7 hours. The teams of gatherers and blowers all had a ½ hour break in each 8 hours. On dayturn it would be 12 to 12-30, in the evening 8 to 8-30 and early morning 4 to 4-30. They finished at 4 pm on Saturday the tank being put on soak until midnight Sunday when it was restarted.

Flattening and Annealing

Glass made in the form of a cylinder had to be flattened to make a sheet of glass, and that sheet of glass has to be annealed and cooled slowly to avoid it breaking up or becoming brittle.

Flat, spread or broad glass as it was known was made from small blown cylinders prior to the 18th Century. After the cylinder was sheared, it was flattened on a flat surface such as a block of marble and then placed in a kiln or in the upper part of a beehive furnace to be annealed.

When the company started making window glass in 1826 it was by

Figure 34. ***Flattener flattening the cylinder on the lagre in the flattening kiln***

the crown process. The tables made by this process were piled in annealing kilns built around the base of the cone that covered the glass furnace. When the kiln was full it was cooled down slowly so that the cooled glass could be taken out and sent to the warehouse for cutting into window panes. In 1790, according to Baynton, the annealing kilns were 12′ long, 10′ wide and 9′ high, with a grate 1′ 1″ wide 4′ 6″ long 8″ deep with an iron door. The piling stones inside were 9′ long 2′ 3″ broad with a hollow 2″ deep and 1′ broad for its full length. The tables were carried into the kiln on a horn which looks like a fork and leaned against a drosser. After a number of tables have been leaned against each other another drosser is put in. In 1850, the annealing kilns were of the same design but much larger. They were then capable of carrying 20 or 30 tables between each drosser and a total of 350 to 400 tables in each kiln.

When the crown process and the making of broad glass was superseded in 1840, by a much improved process for hand blowing larger cylinders for making sheet glass. These cylinders were split with a diamond and flattened. Initially the method used was to place the split cylinder on to a lagre, a sheet of thick glass, in the spreading compartment of the flattening kiln and the sheet spread, Fig. 34. Once

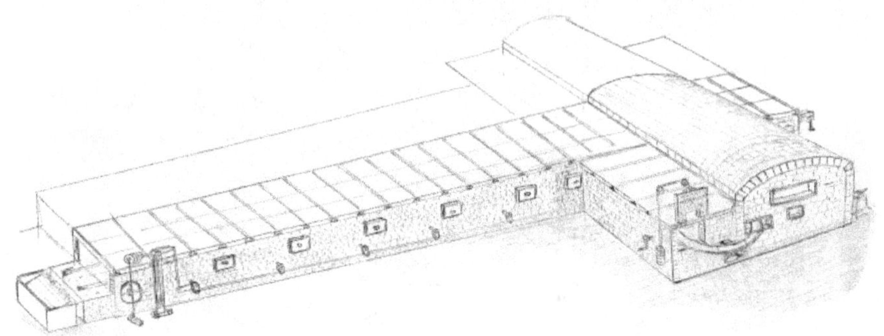

Figure 35. ***A flattening kiln with attached Bievez lifting rail lehr***
It was common practice to build two operating kilns and lehrs side by side
Scale approx. 1:60

this had been done the lagre was pushed through into the cooling chamber where the sheet of glass was removed and piled in the annealing chamber. The lagre was then pulled back and the process repeated. When the annealing chamber was full, it was allowed to cool down slowly before the glass was taken out. A lagre provided the surface on which the cylinder was flattened and it was in continuous use until such time when it needed replacing. This process of flattening gave rise to many faults, such as cockle, sulphur, and dirt in the finished sheet.

The book written by Professor Barker gave a lead on Deacon's patent for the development of an improved flattening kiln with the wheels protruding through the floor which enabled a stone on which the lagre now rested, to be pushed under a tweel into the cooling chamber, where the sheet was taken off with the fourchette and piled in the annealing kiln. The stone and the lagre could then be pulled back into the flattening chamber for another sheet to be flattened. This improved design of flattening kiln did much to do away with the dust problem and by leaving the lagre on the stone did much to improve other faults. It also provided much better control of the temperatures for flattening and cooling. The patent application of 1847 detailed the construction of the lifting tweel between the spreading and the cooling compartments. This tweel was there to allow bent glass, bottles and other ware to pass through. It also detailed

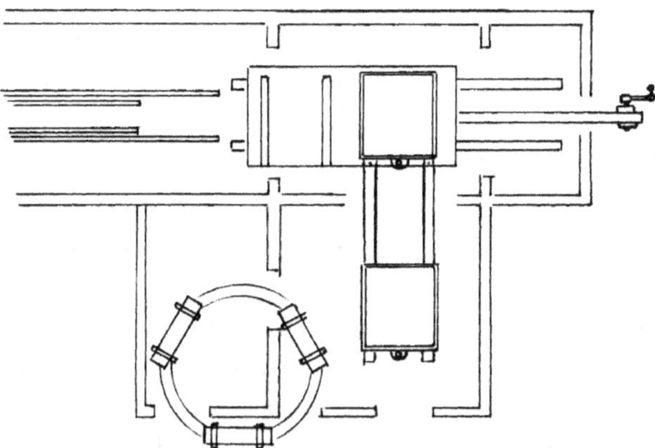

Figure 36. ***A plan view of the flattening chamber showing preheating of the cylinders, the flattening stones, and the carriage at the entrance of the lifting rain lehr***

the construction of the cast iron frame and wheels that enclosed the stone and the runners underneath it. The clearance between the stone and the bed of the kiln was 1/8 inch.

Flattening and annealing glass in the foregoing way was a very slow and intermittent process. It was in vogue right up to 1870 when the continuous Bievez lifting rail lehr was brought into use and the flattening kiln redesigned with two stones for continuous production, Fig. 35.

The way the process worked was as follows. The splitter split the glass cylinders on a specially designed inclined table using a diamond on the end of a cutting lath. The diamond no doubt by Sharrat and Newth had a wooden handle with knob on the top that was used to run the cut. The table was lined with felt and it had a 4″ × 1″ groove down the middle to run the cut and to collect the shards. There was an inspection light to detect any faults. Glass carriers then took the split cylinders and stacked them on the donkeys besides the flattening kiln. They had to carry them in a special way to minimise accidents. In the case of photo glass the split had to be held apart by spring clips. Then it was taken from the donkey by the pusher and put on a saddle or cradle on the dollum. The pusher then wound the dollum round so that the cold cylinder could be warmed, Fig. 36. This automatically placed a warmed cylinder in the flattening chamber for the flattener to place on the lagre. The flattener then flattened the glass on the lagre which was a thick piece of glass on top of the stone treated with gypsum or some other concoction to stop the glass sticking to it. Once the glass has been flattened, the stone which was on wheels, was pushed by the pusher into the cooling chamber. where it was wound down ready for the flattener to put onto the lifting rails of the annealing lehr. There was a lip on the stone to avoid the glass slipping when it was lifted off the stone with the fourchette. At the same time as the stone was wound down, an empty stone was automatically brought into position for pulling into the flattening chamber for the next cylinder to be flattened.

Flattening was carried out in the following way. After the stones have been cleaned at the weekend new lagres were fitted which were 6″ wider than the glass to be flattened. They were treated with gypsum or a solution used by Chance Brothers that was painted on. The flattener using the crappe lifted the warmed cylinder from the dollum cradle on to the stone taking care to place it in the centre. He then turned up the gas to raise the temperature to soften the glass. As it softened, he with great skill opens up the cylinder so that it lies flat on the lagre. The final flattening was done with the pollissoire, which was a block of wood on a long handle soaked in water. The blocks of wood had a hole down the centre so that they could easily be spiked on to the polissoire.

The flattener would be responsible for the heating of the flattening kiln as well as the lehr. He controlled the temperature according to the way the glass softened. The flattening kiln was heated by gas jets, which drew in the combustion air. There was no chimney so the heat from the flames spread throughout the kiln and into the lehr. By this means the cooling chamber and the lehr was provided with an ideal temperature gradient.

The annealing lehr was operated by the pusher who as each flattened sheet was loaded, pulled up the lifting rail and wound it down ready for the next sheet. Then he went to the lehr end to take off a sheet that had been annealed, put it through the washer, on to the drying rack for loading on to a glass barrow before taking into

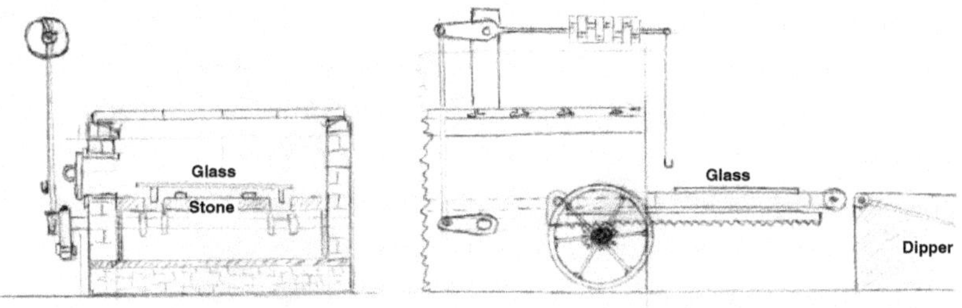

Figure 37. ***A cross section of the lifting rail lehr and a detail drawing of the winding out rack and the dipping tank***

the warehouse for cutting, Fig. 37. The canal water in the dipper was replaced by towns water which was said to be better. Acid was introduced later to remove any sulphur from the glass. Before glass barrows came into use the glass was placed in a wooden crate which was moved by placing it on a two wheeled cranked trolley pushed by a wheeler man and pulled by a puller boy. Steel or cast-iron runners in the floor eased the movement of these trolleys.

The splitters worked day-turn, starting at 6 am with a break from 8 to 8-30, and another at 12 to 12-30 finishing at 5-30 pm. The flatteners and the pushers worked shift but stopped at 3 pm Saturday when the flattener would clear the flattening stones, take off the gas caps on the gas valves, open up the gas damper, light the gas to burn off the tar in the box and the soot in the flue. When this was done the caps and dampers were refitted and the kiln put on soak for the weekend (midnight Sunday). Meantime, the pusher would clear the lehr of any sheets of glass and clean out any cullet that had accumulated on the bed of the lehr.

When Machine Drawn Cylinder glass was made there was a radical redesign of the blown cylinder flattening kiln and lehr, see Figs 38 & 39. This was necessary because the cylinders were much larger and cut up into half rounds, called a shawls. The method of flattening was very much the same and the flattening kiln worked on very similar principles, but the kiln was now fitted with a 20′ turntable to accommodate the much larger stones. Owing to the size of the kiln and the lehr a preheater had to be used to preheat the air to get the

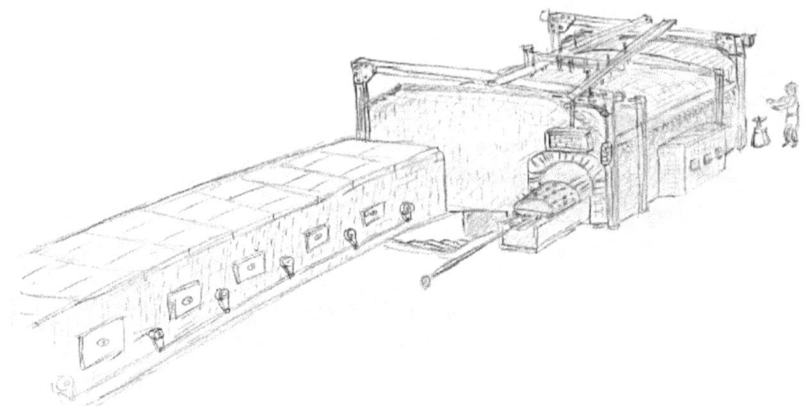

Figure 38. ***Sketch of the flattening kiln and lehr for the shawls***
The preheater is not shown

temperature up to the required level. The shawls were loaded on to a carriage, pushed in and after they had been warmed they were turned over by the flattener on to the stone. There were now four stones on the top of a brick lined turn table or dollum. After each sheet was flattened the turntable was rotated a quarter of a turn to cool it and then after another quarter of a turn transfer it to the annealing lehr.

In detail the procedure was to load a shawl on to the carriage and push it into the kiln for warming, Fig. 40. After it had been warmed the carriage was pushed into kiln proper and the shawl lowered on

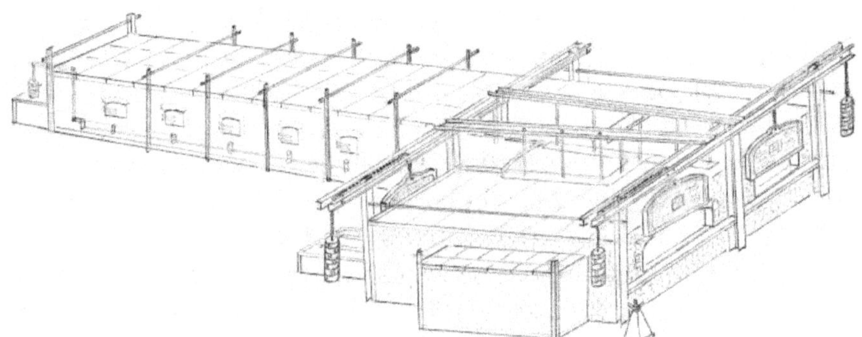

Figure 39. ***Sketch of the flattening kiln and lehr for the shawls***
The preheater is not shown

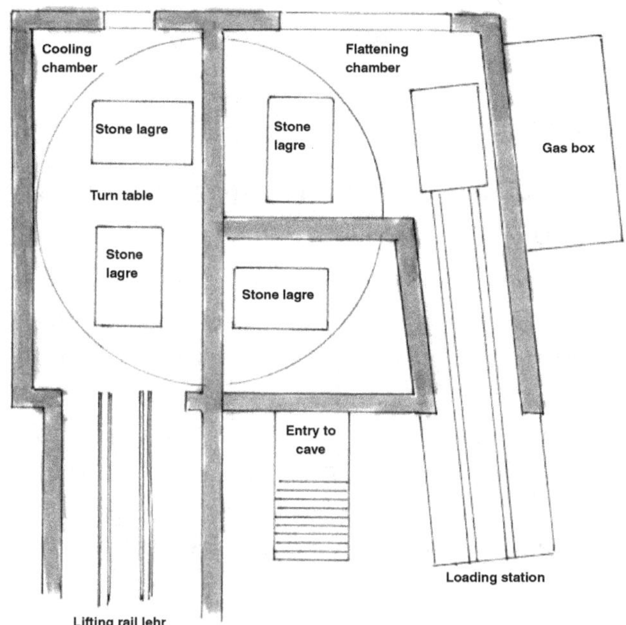

Figure 40. ***A drawing of the flattening chamber showing the 20' turn table and the four stones***

to the iron skids and the carriage pulled back for another shawl. The shawl on the skids was lifted off with the crappe and turned over on to the stone for flattening. The flattener turned up the heat and flattened the shawl at 700°C. The turntable and the lehr was now electrically operated so that when this was done a button was pressed to automatically turn the table and operate the lehr. The transfer of the cooled sheet from the stone in the flattening kiln to the lehr was still done by the flattener using the fourchette, but this automatic link between the flattening kiln and the lehr, signalled the first move towards the use of automatically controlled machines in Sheet Works.

These flattening kilns together with the smaller ones were in use up to 1930. When the flat drawn process took over and they were all dismantled.

CHAPTER 8.

The Divided Continuous Melting and Making Tank 1880 to 1910

When it became evident that the first type of open tank was very difficult to manage and to control, the idea of dividing the tank into two portions, one for melting and the other for making, was tried and adopted. The concept was to have a melting tank separate to that of a gathering or making tank and to build a bridge to connect them together. This enabled the Company to melt and fine the glass and then to pass it into the gathering end. Once the metal was in the gathering end they had control of the temperature, for gathering and working the glass, Figs 41 to 45.

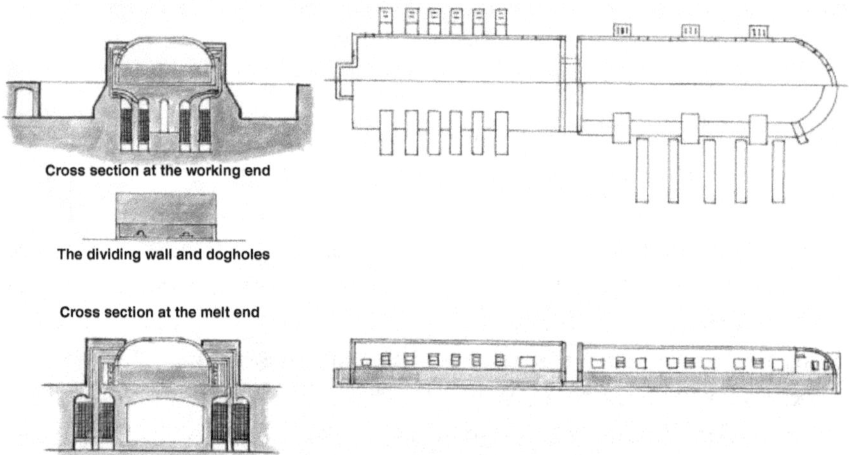

Figure 41. ***Outline drawing of a typical large divided tank showing six stacks to heat the melt end and three specially designed stacks that accommodate the blowing holes and swing pits, for heating the working end. The cross section at the working end shows how the regenerators at that end were designed to clear the swing pits***
Scale approx. 1:120

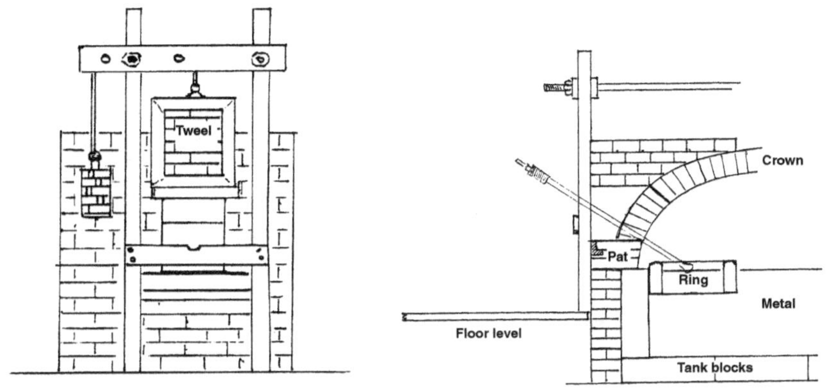

Figure 42. ***A Section through a gathering hole showing a gathering ring and the self wedging tweel***

The Bridge Walls

The first bridge between the two tanks was known as a single bridge. It consisted of a single doghole in the centre, for the tanks were then very narrow. The blocks on either side of the bridge were cut on an angle and the dividing wall was built on top of them, Figs 46 & 47.

As the tanks got wider the single bridge was redesigned and made

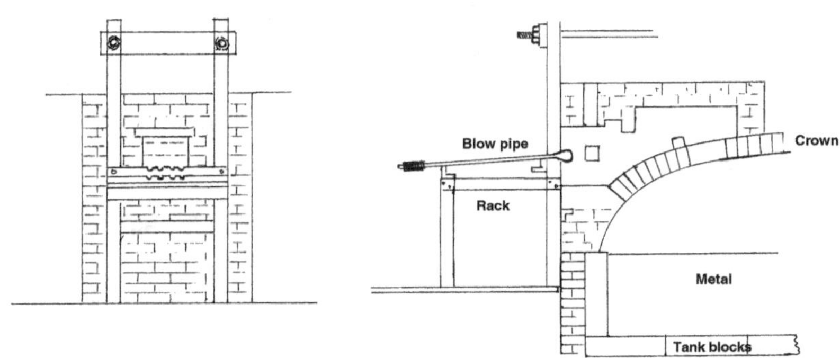

Figure 43. ***A pipe warming hole***

Pipe warming holes were located on both sides of the gathering end of the tank. The pipes were cleaned and placed on the rack ready for use. Beneath the pipe warming hole was the hole for gathering metal on the cordeline tool (not shown)

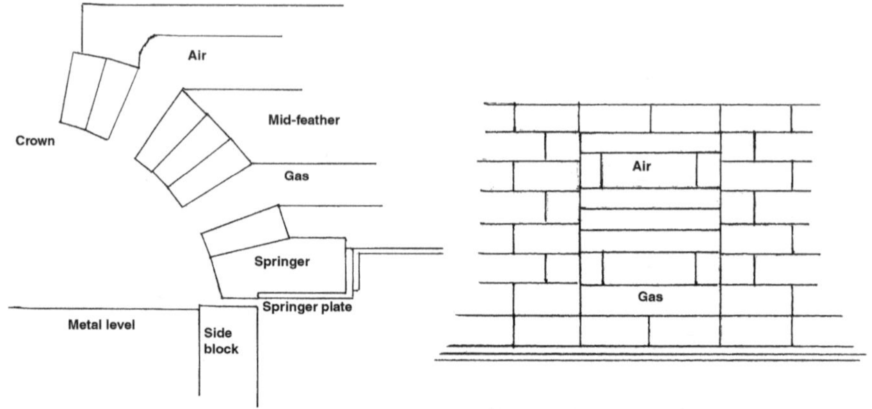

Figure 44. ***Old standard port design***

This was in principle the type of port design that came into use in 1872 and continued in this form right up to 1912. There were variations but the angle of entry of 55° for the air and 18° for the gas remained constant

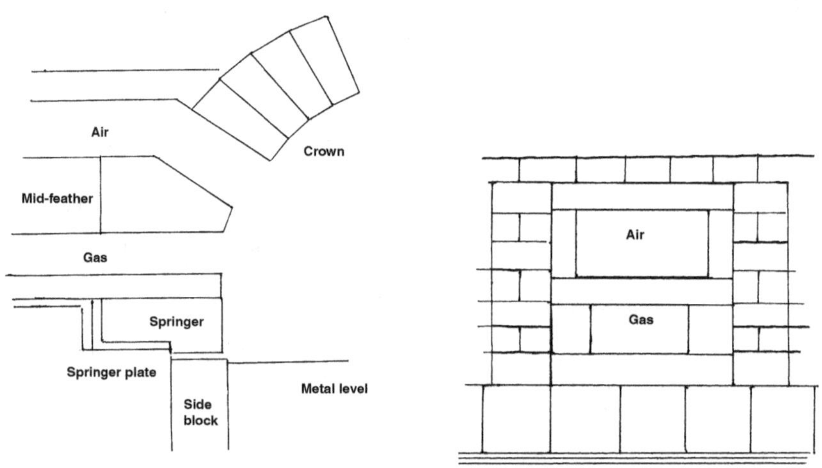

Figure 45. ***The 1920 port design as drawn from memory by Jimmy Hill***

The design brings the gas in at a flat angle but the angle of the air remains at 55°

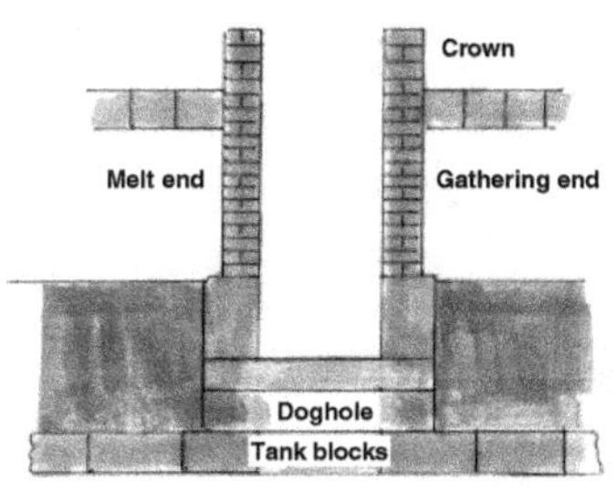

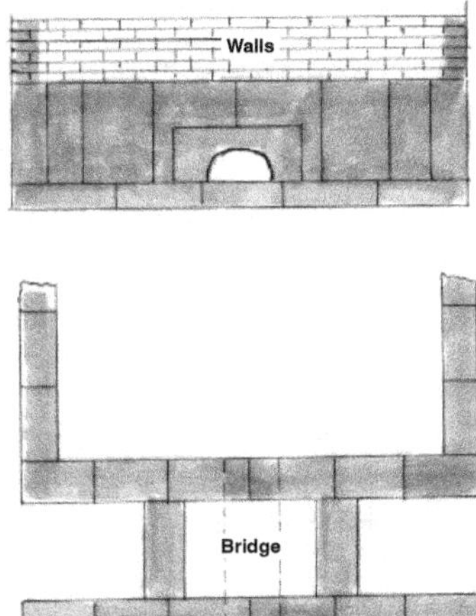

Figure 46. ***A single bridge wall***

into a double bridge. The double bridge had two dogholes and the blocks were laid on the tank bottom. The construction of this type of bridge was quite complicated and it meant that the blocks had to be very carefully cut and shaped if indeed that were possible. It was quite possible that this could be done fairly easily if the K32 block mixture was being used at the time. According to the history of block making this is unlikely, so the only answer may be that blocks were ordered to the shape required (see Tank Drawing Books, PB 95). It is not known why the blocks were cut on an angle, one possibility was

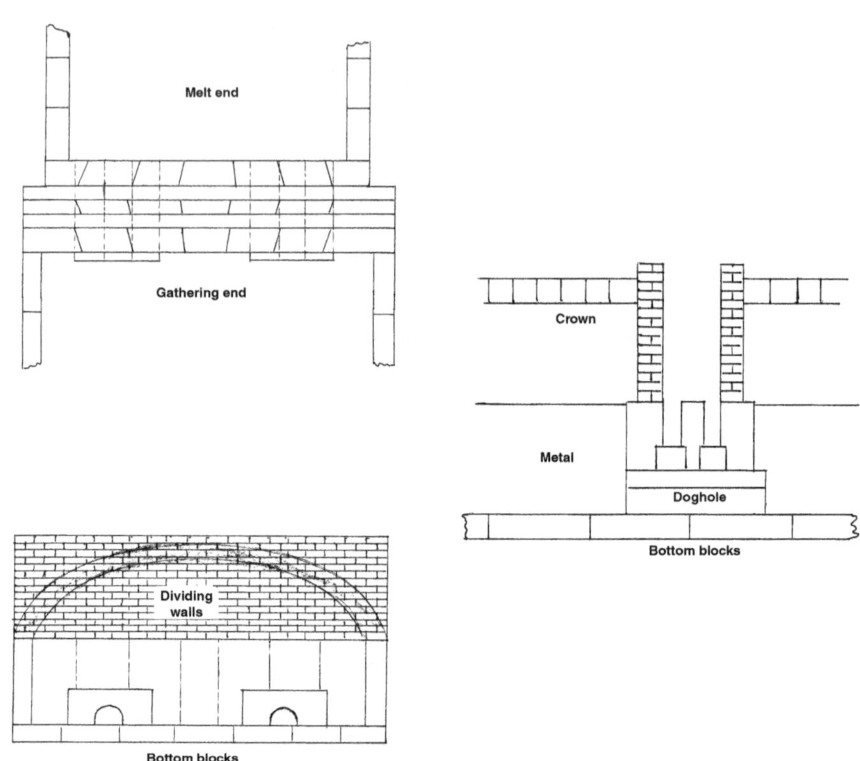

Figure 47. ***A double bridge wall***
The two Dogholes were laid on 4" tiles and the block was 3' 4" long by 45" wide and 16" deep. The Doghole was 1' 3" wide and 9" high. The bridge was built of solid blocks with angled joints but it was not clear why the joints were angled

to increase the width of the joint to encourage the glass to freeze, in what was a very inaccessable area.

Furnace No. 7a, may have been the first tank to be divided, although there was the possibility that some experimental work on a divided tank was done in No. 1 House. This was followed in June 1881 by dividing No. 8b and in July 1881 No. 2b. The divided tank was better for seed, although inclined to ream, and certainly much better for sulphur.

It was this period of development of the continuous melting and making tank from 1873 to 1880, followed by the divided tanks, that founded the prosperity of the company. The concept of the divided tank was quite brilliant and kept secret for many years.

The Operation of a Divided Continuous Melting Tank

The operation of this kind of tank was very similar to that of the open tanks, except that they were larger and there were separate controls over the melting and working ends. As the glass was drawn off by the gatherers the teazers kept filling on cullet and batch to maintain the metal level. The proportions of sand and salts were weighed out by the mixers who were still using shovels for mixing it to make up a batch. It was barrowed into the tank building and emptied on to the filling pocket. It had been reported earlier that frit dust caused trouble in strong winds and at some time later it was noted that at Pilkington's the frit was crumbly. The solution may have been to add water to the frit, and a practice that appeared to be continued in the 1930s when the mixers just pulled the chain on a lavatory cistern to add a set quantity of water.

The filling pocket was an extension of the tank and the practice was, first to fill cullet on to the molten glass and then to fill on the batch to form a lump and to seal the pocket. When the next batch was ready the lump was pushed out into the tank on its raft of cullet. There were about 4–5 lumps floating in the tank in the process of being melted. They were kept at the melt-end by the teazers who positioned them using lump hooks. The flow of the molten glass from the hotspot to the cold filling end helped keep the lumps under control and provided the lumpman or teazer kept them from going over the hotspot they did not go down the tank and upset the fining.

Rings similar to, but larger than those used for gathering from pots, were positioned under each gathering hole in the working end of the tank. These rings were some 2′ 1″ in diameter and 9″ deep. They were fired in what was a coal fired potarch and launched into the working end through the ringhole whilst still at red heat. The temperature of the furnace was controlled by the appearance of the rate of boil on the lumps and at the working end, the viscosity of the glass for gathering.

When the divided tank came into use in 1881 it was reported that the tanks were not "watched carefully enough when there was a change from doing well to doing badly". In 1882 it was noted that the skimmers and men at the melting end could see when the tank was going wrong but could not get it attended to. Apparently the teazers could turn off the steam on the producers just as they liked and the steam was more often off than on, for the slower the producer was worked the less the teazer had to do. In Jan 1883 the ups and downs of tanks showed the want of steady management. There were many arguments about this throughout 1883—"tanks too cold"— "the quality of the teazers"- "the want of a good gaffer". In September 1883 W. W. Pilkington summed it up as follows:- "speaking of quality it was not by any means the tank that was at fault but the way they were mismanaged, the way in which they were not made the best of, one day up, another day down, no two days alike", and this was thoroughly endorsed by H. Taylor. There can be no doubt that all this gave rise to the birth of tank management and control as it is known to this day.

The Birth of Tank Management and Control

From 1883 on there was a tightening up of tank management and control. Later developments showed that the control of the steam on the producers was taken out of the hands of the teazers and left to the producer men. Gaffers and the teazers now controlled the power on the melt by the use of the natural air damper and the length of the flame. The producer men then had to supply sufficient gas to power the tank and if the flame was short they had to put on more fuel and turn up the steam. The same applied to the power on the working end where the temperature had to be maintained to keep the glass at the correct viscosity for gathering.

CHAPTER 9.

The Large Blown Sheet Tanks 1887 to 1929

The divided tanks were very successful. Over the period 1880 to 1887 there was a steady increase in their size and productive capacity. So much so that in May 1885 the advisability of widening the tanks was discussed, but later in July it was decided to put down one tank for ten blowers rather than eight, upon the new ground on the Jubilee site. In 1887, the new No. 9 tank was built and made a good start and in July 1888 No. 10 started with a double bridge, but was only moderate for ream and blibe. No. 9 tank made photographic glass and seemed to be rather better than No. 10. These tanks were a logical development of the small divided tanks for the main change was to increase their size and melting capacity at the melt end, and to accommodate more blowing holes and swing pits between the gas stacks at the working end. Although the small tanks continued to produce glass, they were gradually superseded by the larger tanks, so that by 1890 there were only 13 smaller tanks working, one under the Old Cone, and two large tanks on the Jubilee site.

In 1891–92 No. 11 tank was built inside a large building designed by Medland and Taylor, Architects. It had six blowers on either side with 12 gatherers, and an entirely new thick bottom with thin sides. At that time 1, 2a, 2b, 3a, 3b, 4a, 4b, 5, 6, 7a, 7b, 8a, and 8b. were working.

In June 1893, W. W. Pilkington wanted in principle say, five large tanks including one under repair. It was decided to build them. As regards larger versus smaller sized tanks the figures that he had got out showed very distinctly "the superiority of the former over the latter in cost of fuel, repairs, and almost everything."

In 1894 No. 12 tank was built with a cone and 6 cell kilns.

In 1895 No. 9 tank put out and restarted with double bridge.

In 1897 No. 9 tank put out again and restarted 1898 with maybe five ports.

In 1897 No. 13 tank was built.

In 1904 No. 14 and 15 tanks were built.

It was interesting to note that No. 9 and No. 10 tanks and for that

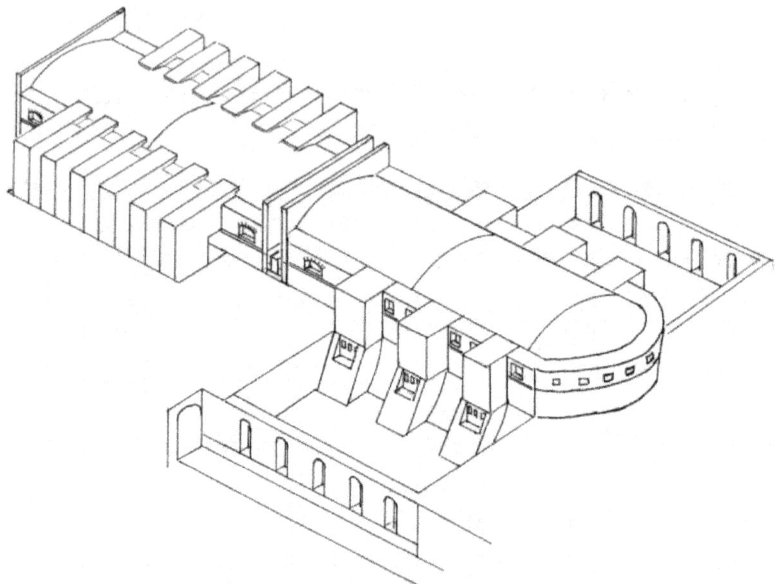

Figure 48. ***A perspective view of a typical large divided tank showing the swing pits and the access tunnels***

matter 11, 12, 13, 14, and 15 tanks still had a cone over their working ends. This cone followed the traditional cone that had been adopted when blowing furnaces came into use for controlling the atmosphere around them, as exemplified in the print of the Works in 1879. The cone and the louvers in the building controlled any draughts that might upset the gathering and blowing, even the large tank buildings in 1936 covering the 1000 ton flat drawn tanks still had a cone like structure over the working end of the tank and louvers in and around the building together with revolving access doors.

It was reported that the average life of the small tanks was now 20 months and for the large tanks 30–50 months for bottoms, sides and bridge. The life of the crowns was also better.

At this stage the output from these new tanks was such that more and more flattening kilns were required. During this period some 40 flattening kilns were built in place of the small tanks, and at the same time developments in Rolled Plate were taking place on the rolling of glass.

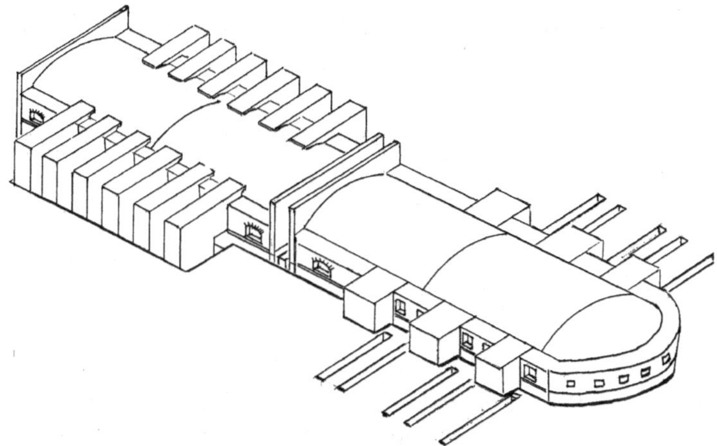

Figure 49. ***A perspective view of a typical large divided tank showing the swing holes at working floor level***

The Excavation and Reconstruction of No. 9 Tank on the Jubilee Site

The site of No. 9 Tank under the Sheet Works dolomite store building was excavated in 1992–93 to reveal the whole of the flue system, the regenerators and the swing pits below the floor of the tank at the working end, Figs 48 & 49. There was little left of the melt end apart from the gas flues and a portion of the gas and air regenerators. The dolomite store building with the cone inside, was largely the same building that housed the working end of the tank, except for some shoring up.

The origin of this type of glass furnace was in 1873 when the first open tank was commissioned using four stacks and possibly a floating bridge, with two gathering holes and four blowing holes in the working end of the tank. In 1881 this type of tank was divided at the bridge by a bridge wall and heating introduced into the working end. Between 1881 and 1887 the size of the tanks increased steadily from two gathering holes to four. At the same time the width and the depth got greater, so that by 1887 when No. 9 was built it had five gathering holes, ten blowing holes, a depth of 3′ with 9″ tank blocks and an internal width of 12′ 6″ with seven small stacks at the melt end and a double bridge. Over its life and up to its demolition in 1929, a period of 30 years the size and productive capacity was steadily increased so that it finished up with six large stacks at the melt end, a depth of 3′ 6″ and an internal width of about 18′.

It was housed in two buildings, the melt end in one, and the working end in the other, with the division wall in between. There was a cone over the blowing end in accordance with the tradition for having cones over the blowing furnaces to control the atmosphere and working conditions. The furnace was fed with gas from eight or ten bell brick producers, one supplying the melt end and the other the working or gathering end. Air was supplied through a natural draught damper and the pull came from a chimney with three eyes. The cave at the melt end was quite large for it housed the two butterfly reversing valves, the pull damper and the entry to the regenerators as well as the two passages down the side of tank used for clearing cullet from the swing holes. The working end also had two butterfly valves for the gas and the air as well as a flue to the chimney. The reversing gear levers for the butterfly valves were located on the furnace floor.

There were two gas flues leading to the working end, one supplied gas to No. 9 and the other to the adjacent No. 10 tank. These gas flues were between the building and the canal and beneath the furnace floor. The furnace floor at the gathering end was at some time or another extended by cantilevering it over the gas flues using concrete for the supports, probably to accommodate an increase in the size of the tank and to continue to give sufficient room for the gatherers to work.

On the drawing of the excavations it has been possible to trace the flue system feeding the regenerators at the working end together with the two butterfly reversing valves, Fig. 50. It has not been possible so far to identify the gas control valve at that end nor the eventual line of the chimney pull flue and its control damper. At the melt end it has been possible to identify the cave, the line of the gas and the air flues, together with the location of the two reversing valves, and the pull flue. The chimney location was well established and there was no doubt that the access tunnel on that side had to be diverted a little to keep clear of it. There are three eyes to the chimney, one eye leads to the pull flue at the melt end, the centre eye may have

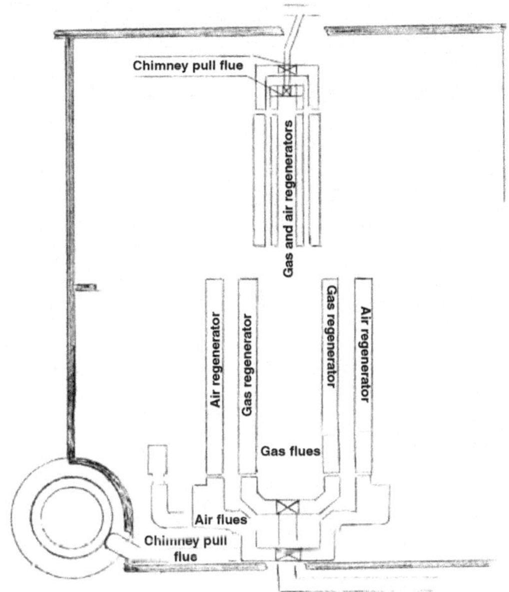

Figure 50. ***Plan of the flue system and the regenerators below No. 9 Tank***

been used as the pull flue to the working ends of both 9 and 10 tanks, the third eye leads to the pull flue at the melt end of No. 10, Fig. 51.

The furnace bottom and side blocks were laid on tiles supported on arches which crossed over from each of the support walls. The melting end crown had about three expansion joints down its length and it was sprung from springer blocks just above block level and built on centres with an end wall at the filling end outside it to allow the crown to expand. The working end crown, also built on centres, was sprung from the blocks except for the rounded end which was sprung from a springer plate to accommodate the gathering holes. The crown itself incorporated ring holes, pipe holes, cordeline holes, and gathering holes.

All the gathering holes were covered by tweels that were lifted by compressed air by the gatherer. The blowing holes too were covered by a shade, operated by compressed air, to keep the heat off the blower.

On the working floor there were the forming blocks, cooling water, chevalliers, and blowing pipes. The skimmers who worked at the ring

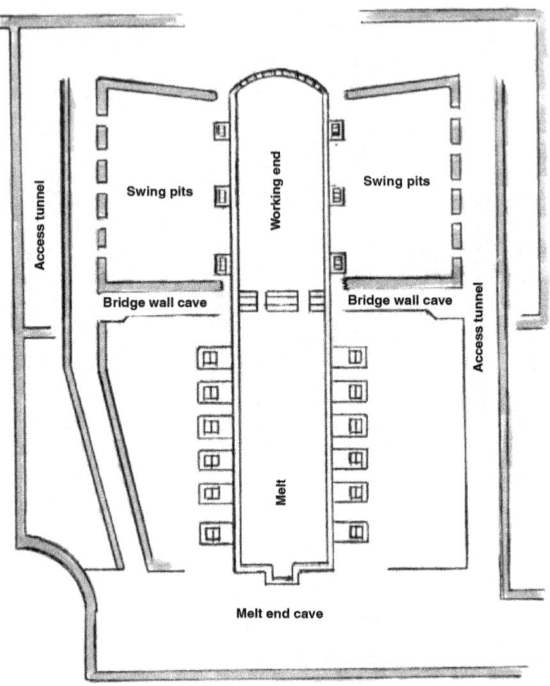

Figure 51. ***Plan of No. 9 Tank showing the cave, the access tunnels, and the swing pits***

holes in the melt end, raked and ladled surface scum, salts and other debris out of the furnace into wheeled ladles, from the front of the bridge wall. There were some jigger or butcher rails round the working end but it is unlikely that there were any at the melt end to carry cullet and frit, for it has to be assumed that barrows were still being used to barrow frit, to fill the pocket. Cullet was removed from the swing holes through the access tunnels down each side of the tank, below floor level, both of which lead into the cave which in turn had an access door leading to a slope up to the yard above.

How Window Glass Was Made in No. 9 Tank

The three ingredients were sand, limestone and saltcake. These ingredients were weighed and mixed in the nearby mixing room. Sand came from the sandfields from around Knowsley, Rainford and Crank, villages to the NW of St. Helens within a radius of about 3 miles. Some of the sand may also have come in from Belgium by boat to Liverpool and delivered to the works by rail. The sand from our own sandfields was by the start of No. 9 in 1887–89, washed at the Mill Lane siding, drained in the railway sand wagons, and sent into the works. In 1901 the company had shares in a Belgium sandfield company and in 1910 bought some 35,000 tons. Nitrecake and saltcake came from local chemical works, but mainly from Cheshire where there were known large deposits of salt. Limestone initially came from local quarries such as the one in Hard Lane, but later the main supply came from North Wales. It was crushed in the works. The other ingredients in the mix were carbon in the form of anthracite and arsenic. The anthracite was to mop up any excess salt and the arsenic to reduce the colour, for the excessive iron in the sand produced glass which was very green in colour.

In the early days the ingredients were mixed by hand, but by the time No. 9 Tank came into operation mixing had become much more sophisticated, so much so that there were bucket elevators, pan mills, weighing machines, and jigger rails. There was no indication how the frit was delivered to the tanks but by all accounts it was barrowed into the filling pocket of the tank. The fourth ingredient if it can be called as such, was the cullet. Cullet is broken glass and pieces left over after the window panes had been cut in the warehouse. It also came from glass broken whilst it was being made and the discards from the ends of the blown cylinders. The passage ways under the tank were for the purpose of collecting the cullet from the swing pits and returning it to the melt end of the tank. On the working floor of the tank cullet was collected from the chevaliers and the skimmers ladles.

The method for filling a tank with frit was first of all to push out the lump already in the pocket into the tank. Then fill the pocket with a layer of cullet so that it formed a raft on which the frit or batch was filled in sufficient quantity to seal the pocket. The lump previously in the pocket now floated on the surface of the molten glass in the tank. The teazers or lumpmen using lump hooks pulled it back against the front wall to avoid it floating down the tank before it had completely melted.

The glass melting furnace or tank was fired by producer gas. There was a bank of inclined grate producers using slack which supplied the gas for melting the batch and another bank which supplied the gas for heating the working end of the tank. These gas producers were the forerunner of the brick producer which supplanted them when the very large 1000 ton tanks were built for by comparison No. 9 was a very small tank with a capacity of only 150–200 tons. The gas for melting the frit was delivered through underground gas flues into the cave under the filling pocket where it was distributed to the regenerators by a butterfly reversing valve. Combustion air was drawn in through another butterfly valve located in the air flues also in the cave. The chimney provided the pull for taking away the burnt gases. In the furnace itself the boss teazer controlled the melt by observing the boil on the lumps. There were no thermometers just the colour of the furnace and the boil. He adjusted the power on the tank by adjusting the size of the opening of the natural draught damper and adjusting the amount of gas through the gas valve to keep a full flame length across the tank. The chimney pull damper controlled the amount of pull on the tank and this was always adjusted to keep a flush on the tank and under no circumstances have so much pull on that air was drawn in through the sight holes or the filling pocket. Every 20–25 minutes the teazers reversed the air and the gas from one side to another. The amount that they filled on was sufficient only to keep the metal level reasonably constant. Each stack was fitted with tile dampers in the gas and air upcast. It was the usual practice to have full power on the first three to four stacks and less power on the fifth and the sixth or seventh. The dampering of the stacks was under the control of the tank manager.

No. 9 Tank being a divided tank the melt end was operated almost quite independently of the working end. The working end had its own supply of gas and air, reversing valves, regenerators, and chimney pull. Basically, the control of the working end heating was the same as that for the meltend but in this case the judgement of the temperature of the glass was quite different. Here it was controlled by the boss gatherer for the gatherers did know whether the metal was too stiff or too fluid for getting the weight, or for gather to be workable. The art of controlling the working end was in adjusting the power to keep the viscosity of the glass for the gather constant.

In the same way as the melt end reversals took place every 20–25 minutes, the power and the pull being adjusted in exactly the same way, although at the working end it was not necessary to adjust the power on each individual stack, only maybe to adjust the gas/air ratio, but it was even more important to maintain the flush.

The tank was provided with sight holes and holes for lump hooks, skimming rakes, and rings. Sight holes were for viewing the flame lengths and the colour of the tank. and the ring hole was for taking out worn rings and putting in new ones. The rings were fired in a kiln and launched into the tank through the ringhole where they were floated down the working end into a position under each gathering hole. The gatherer gathered from the middle of each ring whose main purpose was to keep off the scum in the same way as they were when used for gathering from pots.

The main area of activity was undoubtedly at the working or gathering end where the teams of blowers and gatherers operated. There were five gathering holes servicing ten blowing holes. On the floor there were the forming blocks, a supply of cooling water and a rack for the blowing pipes, which had to be cleaned before being put into the pipe hole. On the tank there were tweels covering the gathering holes that were lifted by compressed air. These tweels were tapered in such a way that when they were lowered they were driven close up to the face of the gathering end wall. At the blowing holes there were the blowing machines over the swing pits and the chevaliers for holding the cylinders for the ends to be cut off. On the side of the tank there were tweels over the blowing holes lifted by compressed air by the blower, and shields to keep the heat off the blower when the tweels were lifted.

The gathering and the blowing of the cylinders was as described by Harry Langtree in Chapter 7 on the Blowing and Making of Compressed Air Blown Sheet.

It was normal to work the tanks all week and leave them on soak over the weekend. The weekend was therefore the time to do maintenance and to burn out the flues. There were two problems in those days, one was the wear on the refractories at the flux line and the other the sooting up of the gas flues. There was always some free salt on the surface of the molten glass and this attacked the tank blocks at what was called the flux line. This meant that the blocks had to be tiled when they were worn through at the flux line, for fluxline water boxes were not introduced until much later. One of the areas for the worst attack by the flux was at the bridge. As it was not easy to tile the blocks in this area the blocks were shielded by rings. It was in this area that runs occurred and this was why there was a walkway above the bridge wall to accommodate fire hoses for spaying on water to freeze the glass.

A feature of the tank was the cone over the working or gathering end. Again this was a relic of the cones over the blowing furnaces and they were necessary to control the atmosphere around the blowing holes, not only for the blowers to be able to work but to avoid toughening the cylinders. Even then with careful control of the louvers it was difficult to blow thick glass without it breaking when it was split. Because of this it was a current practice for one man to hold thick cylinders whilst the splitter ran the cut.

The History of the Site from 1922 to 1929

The tank was demolished in 1929 and it was the subsequent work that went on in the building and its eventual conversion to a dolomite store in 1947 that made it so difficult to interpret the site in 1993–94.

Long before the demolition of No. 9 tank the Company was making enquiries about making of flat drawn glass. by either by the continental Fourcault process or by the utilisation of the Rowarts patents. In 1922 there was a note about "experimental drawn" No. 9 tank and in April 1924 "flat drawn experiments". In 1924 or earlier glass made by the cylinder drawn process was by then supplying the bulk of the trade. No. 13 tank was not in use, No. 11 tank was under repair, and only Nos 12, 14 and 15 tanks were operational. The 1924 valuation sheet confirms that "Experimental Plant being Rebuilt" in No. 9.

A Fourcault machine was never built and it was only in 1928 that the possibility of a Fourcault machine out of the middle of No. 6 tank or even a six machine unit was discussed. It can be reliably assumed that it was the Rowarts process that was experimented with.

As related by Jimmy Hill a four pot furnace was built at the melt end of No. 9 tank using the gas normally supplied to that tank. Later in 1928–29 a kiln was built in the middle of the melt end over the concrete bases shown in the plan of the excavations. This was a glass stretching kiln. There had always been trouble with glass quality arising from the use of a lagre in the flattening kilns. Thought was therefore given to trying to do away with the lagre and one way was to flatten the glass by stretching it. The kiln was built with a sufficiently wide opening to take a shawl. The shawl was suspended from one edge inside the kiln. Once inside a large iron door insulated with asbestos was closed and the kiln brought up to temperature in the hope that the glass would soften and under its own weight flatten itself.

In 1929 it was suggested that No. 10 tank (the one next to No. 9) be converted to an experimental tank using the PPG process and by 1930 it had been rebuilt and the experimental work started. This new tank was a quite different design to that of No. 9. It was an open tank with a floater in the middle and the PPG drawing machine used a shut-off, skim bar, draw bar, as well as "L" blocks, and curtain arches. This clay work had to be fired on site ready for installing in the drawing kiln. The only place for this was in No. 9 tank and two burning kilns were built in there to fire the floaters and the kiln clay work. These kilns weigh about 20 or more tons and this meant building new foundations. These new foundations made a mess of the existing swing holes, flues and passages in that area. Not only that but gas was no longer available from the flues along the side of the canal, so a gas flue was constructed right down the middle of the foundations of No. 9 leading into the valve place where the reversing valves were removed and the gas flue lead into the pull flue. This new claywork needed higher firing temperatures so a preheater using towns gas to preheat the combustion air was installed. This preheater was of the type used on the shawl flattening kilns.

Finally in 1947, the building was converted to a dolomite store. This meant that the walls had to be buttressed and the floor concreted. This conversion caused even greater disruption to the original foundations of the tank.

The History of the Tank from 1920 to 1929

In 1908, No. 8 Tank was built replacing the former 8a, 8b and 8c tanks and brought into production to make sheet glass using the machine drawn cylinder process. Later in 1912, No. 7 tank was built to make glass the same way. Although these tanks together with the blown cylinder tanks were worked right up to 1920, the machine drawn process was steadily superseding the blown cylinder tanks, for after the 1914–18 war, the bulk of the trade was being supplied by 7 and 8 tanks. Research shows that No. 9 Tank was phased out in 1920–21 and that in the period between 1921 and 1929 especially when No. 6 Tank came into production making cylinder drawn glass in 1926. All the other tanks 11, 12, 13, 14, and 15 were phased out except for No. 10 Tank which was used for experiments on rolling and making seed free glass.

In 1919, the Company was aware of the development of the PPG (Pittsburg Plate Glass Co.) process for making flat glass, as well as Fourcault, and the Libby Owen-Ford processes, for as far back as 1912, the glass industry was trying to make flat glass by direct draw. By 1921, there were two processes in operation, Fourcault in Europe and Libby Owen in America which produced flat glass by direct draw but failed to meet our standards for surface quality. They were also aware of the Rowarts Patents and Rowart was visited in Belgium in 1919 and agreement entered into.

Laboratory experiments (Reminiscences J. P. L. Truesdale) had been carried out with syrup, treacle and sealing wax as well as from small pots of molten glass. From 1921 to 1926 pot and tank trials were carried out on No. 9 tank, using the Rowart patents, 182,805 1921 and 212,545 1923–24. When No. 9 tank went out of production, the melt end was converted to a four pot furnace by blanking off the tank and the regenerators just beyond the second stack, Fig. 52 shows the location. Thus, by making the use of the gas and air regenerator at that end, frit could be melted in the pots. The use of pots in the melt end of the tank was a convenient way of obtaining a source of metal at little or no capital expense, whereas a bath extension to the working end of a tank would have involved a cold repair, loss of production and considerable capital expenditure. The techniques for operating a pot furnace were well known and pots could be made

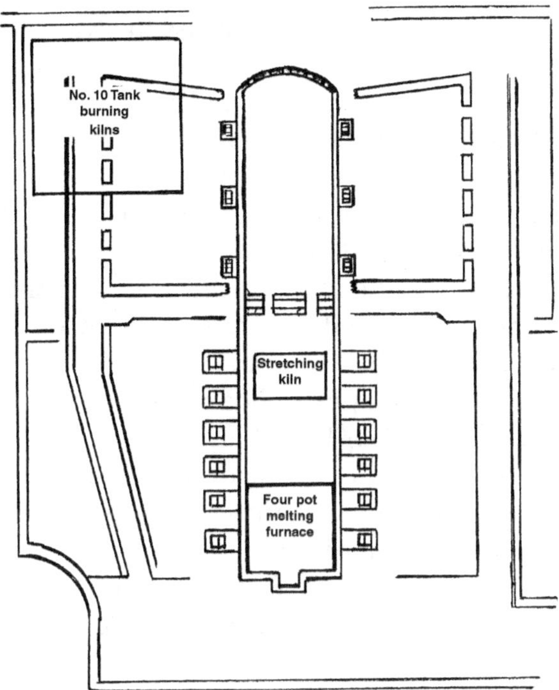

Figure 52. ***Drawing showing the location of the burning kilns for firing the claywork of No. 10 tank, as well as the location of the stretching kiln and the four pot melting furnace***

readily available.

Once the glass in the pots had been founded a pot was taken out and placed in a drawing kiln preheated by towns gas. The earliest experiments were based on the Rowart patents using a strip of wire mesh as a bait drawn upwards in a tower by a series of grips mounted on chains. It produced a sheet of glass 18″ wide and at first 20″ long. It was reported that the longest run was 370′. It was more likely that the 370′ referred to the amount of glass produced on one run, and not the length of the draw. There was evidence of a building between the melt end and the working end buildings of No. 9, shown in an aerial picture postcard of about 1924, that could have been built for housing the tower and the cutting off of the sheet. The quality of the glass produced was very poor.

In 1923, there was an option to purchase the Rowart Patents but there were difficulties in making the process work. In 1924, it was reported that No. 9 was down to be rebuilt as an experimental tank and that we should not abandon the Rowart system and proceed with drawings to give effect to alterations and carry out necessary experiments. at a cost of a few thousand pounds. Despite this they were abandoned in 1925. Mr Cecil however continued with his own experiments in 1926 and he reached the stage of using what was likened to a money box to hold the edges of the sheet. The so-called money box was being preheated in the pot arches when a halt was called to any further development because of the need to conserve energy during the general strike of May 1926.

It would appear that Mr Cecil had got very close to controlling the edges using a Rowart kiln but not infringing the patent. We know that unlike the PPG bowl the money box was made of fireclay. It could have been in, on, or above the surface of the metal but somehow or another the edge had to be cooler than the rest of the metal to avoid it running in to the centre. How close he got we shall never know.

No. 10 Experimental Flat Drawn Tank

In 1919, the Company was aware of the PPG machines, as well as Fourcault, Libby Owen-Ford, and the Rowart process. In 1922 No. 10 tank was adapted to roll a sheet 6′ wide based on the Ford Plate Glass methods. In 1925, No. 10 appeared to be operating normally for the bridge was altered in an attempt to make seedless glass. Between 1925 and 1928 except that the company was getting concerned about the developments of flat drawn elsewhere and our lack of progress, there was a lot of confusion and little information In 1928, the view was expressed that it was essential to get into commercial operation as early as possible for in this year the company did not renew the Rowart patent. Consideration was to be given to Fourcault and LOF. In June 1928, there was a report that LOF was the best mechanically and it was confirmed that "we try one or both at the earliest possible date".

It was also decided to:-

(a) carry out a series of experiments recommended by A. C. Pilkington.

(b) consider lodging an application for compulsory licence for the LOF process.

There was no indication whether or not any of these intentions were carried out. In Feb. 1929, they were considering operating one drawn tank only (No. 6) and purchasing foreign sheet for the other markets. It was agreed that we could not drop out of the sheet trade without endangering our other trades. Finally in 1929, A.C.P. visited America and it was the PPG process that was decided on. In 1930, J. B. Watt visited the USA and it was decided to adopt the PPG process and to alter and equip No. 10 as an experimental tank.

This tank was to a completely new design with a drawing kiln incorporated in the working end. As it was an open tank it needed floaters and the kiln needed claywork in the form of a drawbar, skim bar and shutoff. There was a shortage of space at No. 10 and two claywork burning kilns had to be built on the site of No. 9, Fig. 52 shows the location. Double sets of rings and kiln claywork would have to be fired in case any of the pieces were broken.

The Experimental Glass Stretching Kiln

From the very beginning glass made in the form of a cylinder was split, and then flattened on a lagre or stone. In France and Belgium the flattening operation was called stretching. Stretching meant that a cylinder of glass was stretched out on the stone to form a sheet. So the stretching kiln was not there to stretch glass as was thought but to flatten it.

Flattening in the normal way on a lagre using the crappe and the polissoire could spoil the excellent surface of the glass. The stretching kiln was designed to flatten shawls without marking or spoiling the surface of the glass. It could have been, an adaptation of the very early flattening kilns used in 1840–50 which used a warming chamber, a flattening chamber, and an annealing chamber.

The warming chamber in the original design was redesigned to accommodate a shawl. The shawl was suspended by tongs from a butcher rail and pushed into the chamber. An iron door lined with asbestos was closed and the chamber heated to the softening point of the glass. The shawl under its own weight would then start to

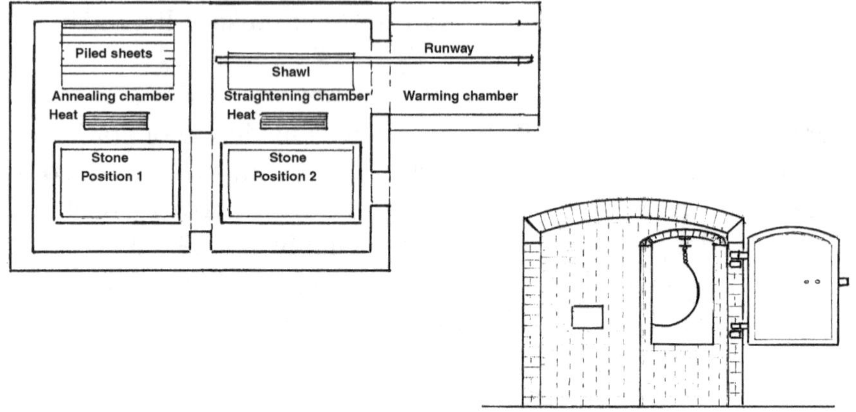

Figure 53. *The stretching kiln*

The plan shows the internal layout of the stones, the heating jets and the position of the piled sheets in the annealing chamber. The end view shows the oven door and the shawl suspended on the jigger rail Scale approx. 1:60

flatten. When it had stretched out as far as it would go, the chamber was cooled and the sheet of glass transferred on to the flattening stone, where it would be reheated to complete the flattening process. Finally, the stone was pushed into the annealing chamber for the glass to be cooled and piled. When about 10 to 15 sheets had been piled the kiln, it was closed and cooled down slowly to the anneal the sheets of glass and to examine them for faults, Fig. 53.

To carry out these operations, the kiln would need two independent gas jets, one for the flattening chamber and the other for the annealing chamber. The dimensions of the stretching kiln would be about a minimum of 20′ × 14′ × 14′. It would be operated in the following manner:-

1. Hang the shawl on chains from the rail using a clamp.
2. Run the shawl into the kiln on the rail and close the door.
3. Heat the chamber and allow the glass to stretch.
4. Cool it a little and then swing the 70% flattened shawl over and on to the stone.
5. Reheat the chamber to complete the flattening process.
6. Push the stone through into the hot annealing chamber.

7. Pile the sheet on to the back of the kiln and pull back the stone. It was not easy to understand why such experiments were carried out at such a late stage and only a year or so before starting No. 10 and the experimental PPG machine. It may be that with the heavy investment in machine blown cylinder production, it was a final effort to make sure that there was no alternative.

Gas Producers

In 1873 there was the mention of "boilers in mixing room", and in 1879 producer blowers were said to reduce the found by 2 hours. There was no indication when blowers were used with the steam on the producers to produce water gas, but there was evidence that the inclined grate producer reigned supreme right up to 1889 and that they were operated in banks of possibly three or four or more. However, with the development and installation of the larger tanks there must have been a need to increase the supply of gas to the furnaces. By 1901, (see tank drawing books), there were twin blast pipes running down the bottom of what looked like an eight bell brick producer. Steam was injected into the blast pipe through a simple bent nozzle to draw in the air. It was therefore logical to assume that the concept of a brick producer was by joining the inclined producers together, as illustrated by Fig. 54, and comparing this to a drawing of a cross section of a 24 bell producer.

In 1924, No. 13 producer had 10 bells, 14 and 15 also had 10 bells, rolled plate had two 8 bell, and No. 10 producer one 10 bell and one 12 bell. By 1930 the larger 20, 24 or 26 bell producer equipped with two venturi type steam/air injectors nozzles, coal bunkers and a wagon tippler, must have been developed for the large tanks. The brick producer was an entirely Pilkington development which at the time and right up to the 1950s stood up to all competition including the mechanical producers. Mechanical producers finally won mainly due to the reduced labour cost and reduced effect on the environment.

It was interesting to see that in 1886 under the Public Health Act of 1875, the gas producers at Plate Works were causing a nuisance in the area At Sheet Works the gas from the producers did at times

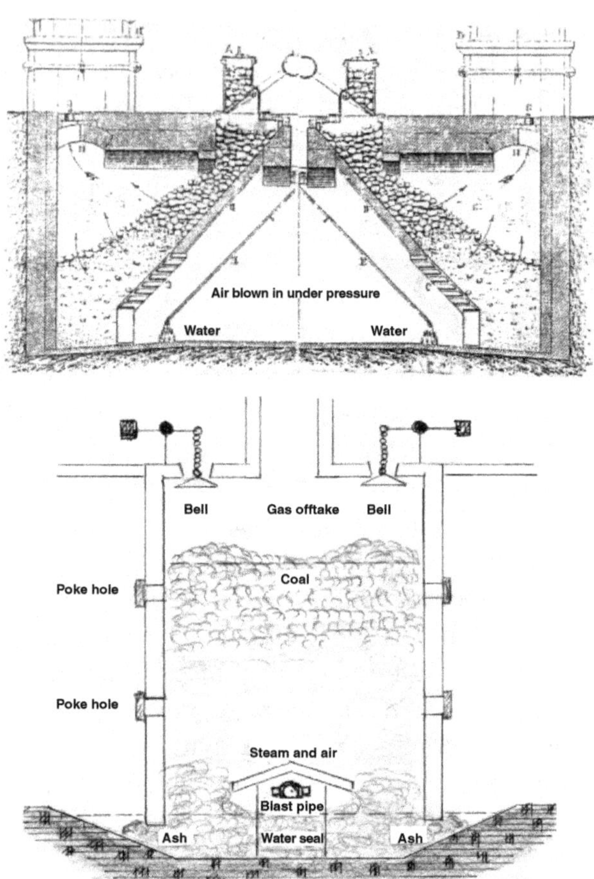

Figure 54. ***The illustration shows how two inclined grate gas producers could have been placed side by side to form a central chamber for providing air and water vapour. This could have been the first step towards the concept and design of the Pilkington Brick Gas Producer which had a blast pipe running down the centre and a water seal for removing the ashes***

cause a pall of smoky gas in Bridge St. It was quite probable that in neither case was it from the producers but from the burn-outs.

CHAPTER 10.

The Machine Drawn Cylinder Tanks

Up to about 1908 all the glass that was being produced was from the blown tanks. In 1903, developments started on the process for drawing cylinders by machines. In 1904, AWG started making glass in America by machine, and the company went to see it working. The first machine drawn cylinder tanks were built as divided tanks and No. 8 tank was built in 1908 for the development of this process. By 1910, No. 8 tank was working with 10 drawing machines, drawing cylinders 18″ diameter, 168″ long and later 32″ × 420″ using redesigned flattening kilns. In 1912, No. 7 tank was built with eight drawing machines. After 1912, new tanks and any rebuilt tanks were no longer divided by a bridge wall, but were built as open tanks. Even though initially, they were large compared to the blown tanks they were, as time went on steadily being widened 2′ at a time and so that eventually, No. 5 tank in Rolled Plate, making cast glass, was accredited the biggest glass melting unit in the world. No. 6 tank (Fig. 55) was built in 1924 on the colliery site with a layout for 10 machines, 10 flattening kilns, a mixing room, 12 pot kilns, and 12 take down machines. The capacity of the blown tanks was in the region of 120 to 150 tons whereas the capacity of the new tanks was now in the region of 800 to 1000 tons. The design was quite different, instead of springing the crown from the side block springer plate, it was now sprung separately from a springer plate on the buckstays. There was no longer any bridge wall, and water boxes were used on the flux lines on top of the melt end blocks. Blowers replaced natural draught for the combustion air. Forter valves, whose design probably came from America, replaced the butterfly valves. Floaters were used across the middle of the tank but not a flying arch for the temperature at the working end would need to be fairly high to allow for the loss of temperature when the glass was ladled into the reversible pots.

This process operated together with the blown tanks right up to 1929. During this period there were experiments with the drawing of a sheet of glass. There was some consideration given to the possible adoption of the Fourcault process for drawing sheet glass. Finally it was the PPG process that was adopted and experiments started on No. 10 tank in 1929–30. These were quite successful and in due course and flat drawn tanks replaced the machine drawn cylinder tanks as well as the blown tanks.

Firing up a Large Tank to Working Temperature

Glass makers did not have any temperature measuring devices such as thermometers or thermo couples in the early days of glass making. How then did they burn clay pots or heat up their furnaces from cold? Burning clay or firing pots had to be very carefully controlled, first of all to drive off any moisture left after six months of drying and then to transform the clay into a refractory. In the early days

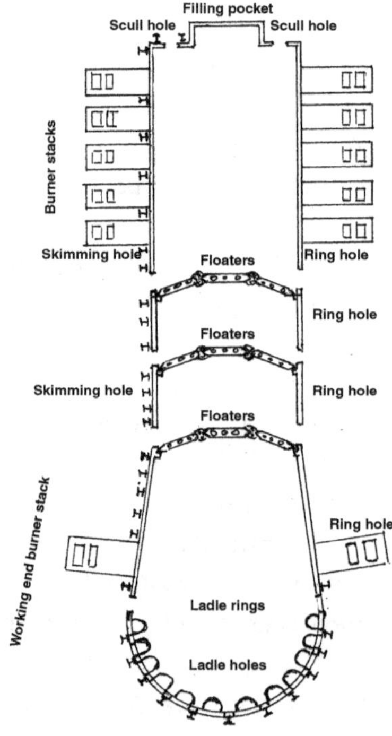

Figure 55. ***An outline plan of No. 6 Tank using three sets of five piece floaters, working end heating and twelve ladle holes***
Scale approx. 1:240

furnaces built of clay brick were heated by coal fires and brought up to temperature fairly slowly. However, when gas firing and silica brick came into use much greater care was needed. Thermocouples were only in their infancy in 1873 and optical pyrometers in 1875. They only appeared to come into real use in the 1900s. Before that glass and clay makers had to bring up their furnaces and burn their clay by sheer experience from sight and feel. It was in fact recorded that Mr Cecil in 1901 brought in a temperature recording pyrometer and in 1905 advised getting an Ishtale pyrometer to record different heats. According to a note left by Mr Cecil and recorded by Mr David it was he who introduced temperature measurement and was accused of not trusting his workmen. Harry Langtree retold the story of E. B. Le-Mare, who, despite all the forebodings of the kiln man, put a thermocouple in a kiln for annealing cells. He was told that "the kiln was not right" and that was indeed right for all the cases at the back were broken.

It was and still is well known that you cannot gas a cold furnace, so furnaces had to be brought up to temperature either by coal firing or by cold gas before gassing. The early gas fired pot furnaces were brought up with cold gas. When they were replaced by tanks, the tanks were brought up using coal fires around their perimeter. Later in the 1950s, the tanks were brought up using town gas fired blowers. At all times the chimney pull had to be adjusted to allow the tank to be under pressure to stop any cold air being drawn in. When using coal fires the standard piece of equipment around the tank was a stick of wood which smouldered when inserted in a sight hole. If the smoke was drawn into the tank there was too much pull. In order not to crack the blocks and to allow the bricklayers to control the rise of the crown as it expanded tanks were heated up very slowly. It would take up to two to three weeks to get them up to gassing temperature.

The Operation of Large Machine Drawn Cylinder Glass Tanks

The process of melting the glass in what were now very large tanks was much the same as that for the blown tanks, except that with the passage of time developments and improvements had taken place.

Division walls were replaced by floaters in the middle of the tank. These floaters were fired in the ring arches and they floated on the metal in the middle of the tank. Floaters on the surface of the metal were there to hold back any dirty metal and the surface flow. The story by Jimmy Hill was that when the tank was built, the floaters were laid in on the tank bottom and brought up with the tank in exactly the same way as the rings were in the pots. Unfortunately, when the tank was filled the floaters stuck to the bottom of the tank. The mixing rooms were now equipped with much more sophisticated pan-mill mixers and the batch was delivered to the tank in bogies on a butcher rail. Lump filling was still in use, with the teazers or lumpmen controlling the lumps and the reversals. Although they controlled the flame length by operating the gas valve, the tank manager controlled the power on the tank by adjusting the amount of air delivered by the Rootes Blowers which had replaced natural draught. Forter valves, with an incorporated water seal, were now in use for they replaced the old iron butterfly valves which could no longer stand up to the heat. Temperature control was by thermocouples in the crown and the metal level was controlled by a metal level detector.

A Summary of the Design and Operation of No. 6 Machine Drawn Cylinder Tank (G. McOnie's Diary, Fig. 56)

No. 6 Tank was 137′ 8″ long and 30′ 4″ wide, 4′ deep at the melt end and 3′ 3″ deep at the working end. It held 1115 tons of glass with cast iron water boxes on the side blocks at the melt. The sides and crown above metal level were of silica. It might have been a relieving arch crown for it was thought that continuous sprung crowns came in later. The filling pocket was 16′ 6″ long by 4′ wide with two scull holes on either side. The gas supply came from No. 6 brick producer with the flattening brick producer as standby for burn outs. The composition of the gas was CO_2 5.6%, CO 24.4%, CH_4 4.8%, H 15%, N 50.2%. The exhaust gas was CO_2 17.18%, O_2 2%, CO nil.

There were two Forter Valves, one for each producer. At the working end there were 12 ladle holes with pneumatically operated tweels and of course 12 ladle rings. The glass analysis was Si 73.1, CaO 14.3, Al-Fe 1.1, SO_3 0.7, MgO 0.3, Na_2O 10.5. It was filled every hour from

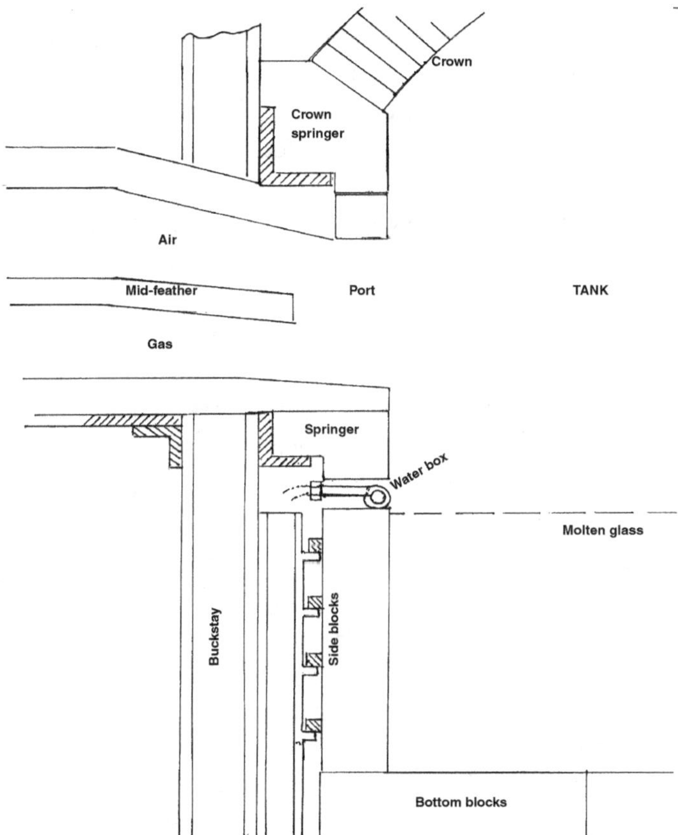

Figure 56. ***A section showing the air and gas port as used in the large 1.200 ton tanks as well as the independant side wall and crown springers. Also shown is the flux line water box and the metal bracing for the side blocks***
Scale approx. 1:16

line bogies, 2 of cullet and 1.75 to 2 of frit. The lump was pushed out every two hours and its melting time was 4 to 5 hours. There were three sets of five piece floaters in the middle of the tank and no flying arch. Dipping rods were used by the tank managers to ascertain the viscosity of the metal. Temperatures were measured by crown couples and the side wall temperatures by radiation pyrometer. The gas regenerators were 12′ 3″ high and 4′ 8″ wide. The air regenerators were a bit larger 13′ high and 6′ 2″ wide. It took two weeks to bring up the tank from cold to gassing temperature and a further week to bring it up to glass making temperature.

The glass was ladled out of the tank by an electrically operated ladle truck. The ladle went in, rotated clockwise to scoop up the metal, and emptied into the pot from the other side. Any chilled metal in the ring had to be cleared and any metal left in the ladle was emptied into the scull ladle which was taken on a butcher rail back to the melt end. Afterwards the ladle was cooled ready for the next fill. One ladle truck served six machines.

Around the working end of the tank there were 12 reversible pots (Fig. 57) under 12 cylinder drawing machines. The pot would be filled and after a cylinder had been drawn out of it, it would be reversed in order to empty it and to bring the pot on the reverse side uppermost ready for filling. Each reversible pot sat on top of a kiln heated by producer gas. It was interesting to note that only the air, not the gas, was preheated by a regenerator. It was possible that there were two sets of air regenerators, one on each set of 6 pots using twin flues and reversal valves. The pot kiln was side fired so that the sculls from the pot, when it was reversed, could go down the centre into a bogie and returned to the melt end. There were two half plates which closed around the pot to seal the top of the kiln. The fire clay pots had a life of about 12 days, but on the odd occasion going for a maximum of 40 days.

The drawing machine consisted of a conical bait that was clipped into a frame. This frame was raised and lowered between two guide rails by a rope on a drum winch. Compressed air was blown into the bait through a telescopic tube. The procedure was to lower the bait into the metal in the pot, hold it there for a few seconds, and then raise it gradually to form the neck by expanding the diameter from 18″ to 40″. Once the neck had solidified at 40″ the draw would commence at a speed and air pressure consistent with the substance that was being made at the time. When the correct length had been drawn, the draw was speeded up and the blow increased to thin off the bottom for wetting off. After that a loop was put round the cylinder to lower it on to the chevalier, where it was cut into sections by wrapping a copper band round it, heating the band and wetting

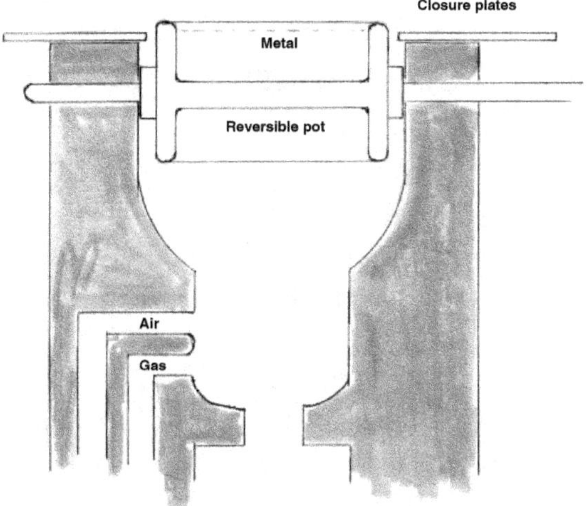

Figure 57. ***A section through a reversible pot kiln***
Scale approx. 1:25

off. The sections were then split into two halves called shawls, which were then taken to the flattening kilns for flattening.

CHAPTER 11.

The Flat Drawn Sheet Glass Tanks for Making Window Glass 1930 to 1980

The advent of the machine cylinder drawn tanks lead to the phasing out of the compressed air hand blown cylinder tanks such as No. 9, 10, 11 12, 13, 14 and 15 towards the end of the 1920s. In 1924, there were experiments in drawing glass as a sheet instead of a cylinder. About this time it was known that flat glass could be made using the Fourcault process. In 1928, the possibility of installing a Fourcault machine in the middle of No. 6 tank or indeed a six machine Fourcault unit was considered. At about the same time it was decided to convert No. 10 tank and to rebuild it as a flat drawn experimental tank using the PPG process that had been developed in America. This was a small open tank with a flying arch, floaters and a single PPG 72" drawing machine, using edge bowls. The drawing machine was built in the cone that existed at the time over the drawing end of the old No. 10, and the claywork was fired in the kilns built on the No. 9 tank site. No glass was made for three months and all told it took about a year to fully develop, so that by November 1929 the PPG process was adopted.

Once the PPG process had been formally adopted for making sheet glass, things moved on very rapidly. In 1931, No. 7 tank was converted to flat drawn and by 1931 was being operated successfully. In May 1932, No. 6 tank was rebuilt as a Flat Drawn Tank and in January 1933–34 a completely new No. 9 was built, on a new site next to No. 7 tank. The old No. 8 machine drawn cylinder tank had by this time been closed down and left derelict.

The fundamental difference between these new tanks and the ones for making cylinders was that the company now had a tank and a machine that produced a sheet of glass ready for cutting up into window glass and not as a cylinder which had to be split, flattened and annealed in a lehr. These three new tanks produced all the glass that was required so that by 1934 all the old tanks and flattening kilns had been dismantled.

No. 6 Tank was equipped with one 72″, one 84″ and three 94″ machines, No. 7 had four 94″ machines, and No. 9 one 124″, one 132″ and two 94″ machines. These widths represented the widths of the sheet of glass as it was drawn. The speed of draw for 24 oz was 30″ per minute and for drawing 18 oz a faster speed of about 46″ per minute. For the thicker substances such as 32 oz, and ¼″ the speeds were slower. These machines made a range of sheet glass from 9 oz photo glass to 11 oz, 18 oz, 24 oz, 32 oz, 4.1 mm and ¼″. There was an attempt to make ½″.

A Flat Drawn Tanks were built with S and G blocks on the bottom and the sides to a depth of 4′. They were about 30′ in width and 100′ to 120′ in length. and held from 1000 to 1200 tons of molten glass. The flux line was at the melt end protected by flux line water boxes and the blocks at that end and down to the middle cooled by fan air. Each tank had a large filling pocket and it was filled by bottom opening canisters mounted on a butcher rail. The cullet canisters were wheeled in from the machine end and the frit canisters from the mixing room. All the superstructure was in silica brick, the crowns being either continuous sprung or relieving arch, with expansion joints across the tank every 20′. The spring of the crowns was supported by buckstaves which also held the springer plates for the sidewall springers. Crowns at this point of time were not insulated although the working end crown in No. 9 was insulated with what was called sawdust brick. Sawdust brick was a loosely bonded brick of vermiculite. There were generally five stacks on either side with one set of rings in the middle floating on the surface of the glass (Fig. 58) and a flying arch that shielded the intense heat at the melt end from the glass at the working end. In the middle of the tank were the skimming pockets and a large hole in the side of the tank for pulling out the worn floaters and putting in new ones. When new or after a cold repair they were brought up to temperature over a period of about three weeks using coal fires placed all round the tank. The buckstaves were let out to accommodate the expansion of the crown and the longitudinal expansion was taken up by the expansion joints. Bottom and side block expansions were not very great the blocks being mainly of alumina and what expansion there was,

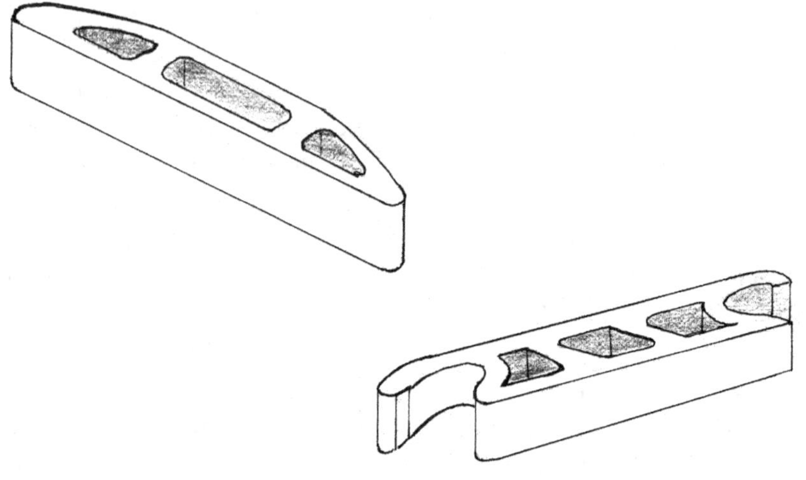

Figure 58. ***Rings or floaters***

Three piece or five piece sets of floaters were floated on the surface of the metal in the middle of a tank. Their main purpose was to divert any scum or or unmelted material into the skimming pockets for raking in and ladling out by the skimmer. The ones illustrated are from a three piece set consisting of an anchor ring and two side rings. Although they are really floaters and not rings, the name ring persisted long after rings were no longer used. Even to this day they were fired in a Ring Arch, and the process of replacing them in a tank was known as "ringing the tank"

was allowed for by the expansion of the iron work holding them in place. Once the tank was up to more than red heat and most of the expansions had taken place the tank was gassed and brought up to working temperature. A new tank was filled with cullet.

These new flat drawn tanks dwarfed the earlier blown tanks above ground but it was not generally realised that there was as much room below ground as there was above to accommodate the regenerators, as well as the reversing systems, the air cooling, and the gas and air flues. Each tank was fed with gas from a brick producer. The size of the producer measured by the number of bells depended on the size and productive capacity of the tank. A small producer had 16 bells and a larger one 24 bells. Coal was fed into the producer through these bells. There were three levels, the bottom for ashing and poking, the middle for poking and breaking up the clinker, and the top for the coal feeds. The base of the producer was in a water seal and the ashes taken out through the water, put in a barrow and tipped into bucket elevator. There were two gas offtakes leading into the main gas flue to the tank. A wagon tippler emptied wagon loads of coal into a bucket elevator which filled an overhead hopper. The producer men drew off coal from this hopper into line bogies which were wheeled over the bells and discharged into the producer. At each and every charge a plume of gas rose into the air and there were always small leakages of gas from the crevices of the brickwork, which everyone got used to.

There were two gas mains into the cave leading into two Forter valves. one main was for the permanent supply of gas and the other was from a spare producer which supplied gas whilst the main flues were being burnt out. A burnout took place when the main producer was taken off line to set the flues alight in order to clear them of the accumulated soot and tar. Combustion air was supplied by a Rootes blower through a reversing valve in the cave which was over the top of the chimney pull flue. The complicated system of gas, air and pull flues fed gas and air to the regenerators whilst the pull flues took away the burnt gas. The regenerators supported the melt end of the tank. They were built with clay or alumina brick and bound with railway metal. Inside were the network of stuffing brick 7″ × 3″ which over a period of 6–8 months needed renewing for the top bricks were melted down and the bottom bricks choked by dust and dirt carried over by the flames. The stuffing was renewed at what was called a regenerator repair. This kind of repair was carried out hot. Water was used to cool the brickwork to just below redheat, apart from that, no time was wasted getting the stuffing out and replacing it with new brick. The regenerator arches and the upcasts were such that there was an outer and an inner skin, so that if the inner arch and the upcast needed repair it could be done by rebuilding them from the inside. The bricklayers were very skilled in centring and placing the key bricks working from beneath an arch. The saddle tile supported the stuffing above the flues and it was amazing how they stood up not only to the heat but to the cooling water poured on to them during a hot repair.

Each tank had its own mixing room. The raw materials were delivered either by rail or road, discharged into hoppers and elevated into storage bins. The mixers drew off the materials into a bogie to the exact weight laid down for the recipe to make window glass. When all the materials had been weighed out they were put into a mixer, a little water added and thoroughly mixed, before being sent into the tank for filling on to the pocket.

Operating a Flat Drawn Tank

They were operated in much the same way as that described in the operation of a blown or cylinder drawn tank. The frit on a raft of cullet was formed as a lump in the filling pocket in such a way that it reached the filling pocket arch and sealed the pocket off from the tank. When the next batch was ready for filling on, the lump was pushed out into the tank where it floated on the metal. At any point in time there would be at least five lumps on the surface in various stages of melting. The lumpmen or teazers were responsible for placing the lumps with lump hooks and pulling them back against the front wall so that they did not float down the tank. They reversed the gas and the air every 20 minutes by throwing over the Forter valve and the air reversing valve. The tank manager controlled the power on the tank by increasing or decreasing the r.p.m. of the Rootes blower, leaving the teazers to adjust the gas control valve to maintain a constant flame length. The power on each stack was adjusted by stack dampers in the upcasts. Normally the first three stacks were fully open to maintain the hot spot between the third and fourth stacks. The fourth and fifth stacks would be dampered down a little to keep a steady temperature in the middle of the tank and on the melt side of the flying arch for fining the metal. The working end of the tank was held at a lower temperature and fine adjustments to the temperature were by means of a fan which blew cold air into it. A metal level measuring device was used to keep the metal level as constant as possible. The temperature of the tank was measured by thermocouples in the crown and the sidewall temperatures by a radiation pyrometer.

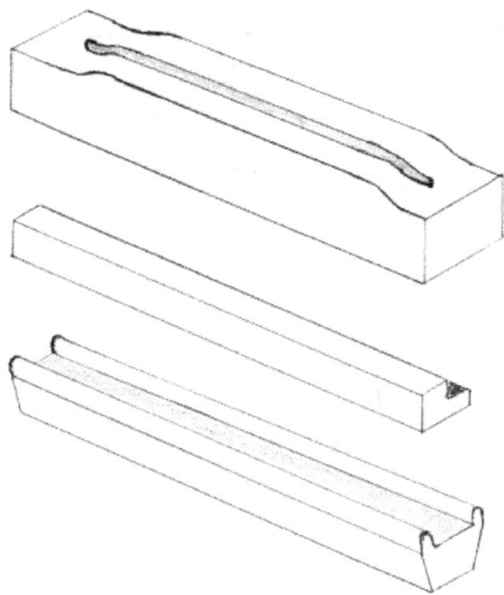

Figure 59. ***Kiln claywork***

The bottom claywork, in or below the metal, in a flat glass drawing machine consisted of a drawbar, skimbar, and shut-off. The drawbar was submerged below the metal by two depressor plugs, one at each end. Its purpose was to stabilise the draw of sheet across the bath. Skimbars were used to skim the surface metal for surface faults, and the shut-off was to seal the drawing kiln or bath from the tank. It was depressed below the surface of the metal by a shut-off tweel

The Flat Glass Drawing Machines

There were four or five vertical glass drawing machines on each tank. Each machine drew glass from the bath of molten glass beneath it. The bath was an integral part of the working end, but sealed off from the tank by a shutoff and shut off tweel. Sulphur that had plagued glass making from the very earliest days was now a thing of the past. Drawing glass by the machine drawn cylinder process had helped to reduce sulphur when the glass was ladled out of the tank but there had still been problems with sulphur in the flattening kilns.

The flat drawn kiln or bath consisted of the shutoff block, a skim bar, and a draw bar, Fig. 59. The draw bar was held down by the depressor arms on the breast block and the skim bar held in place by lugs on the side blocks. The bath was covered by a curtain arch, two "L" blocks and a front arch. Under the front arch were two holes for the gas burners which were used to keep up the temperature in the front of the kiln. Between the "L" blocks and above the bath was the drawing tower. It was about 30′ in height and full of asbestos rollers which drew the sheet of glass upwards. This tower not only drew the glass upward but also acted as an annealing kiln, so that for the first time the company had a fully integrated process for making sheet or window glass.

Operating a Flat Drawn Machine

After the machine had been set-up with the drawbar, skim-bar, curtain arches, and ell-blocks in place two water coolers were slung across the kiln. These coolers were 10″ deep, later on 12″, and still later 14″, and the width of the sheet in length. All the gas burners in the tower were turned on to heat it. The rollers in the drawing tower were reversed and the bait sent down. The bait was an iron frame work with wire coils on the bottom which when dipped in the molten glass, pulled up a sheet of glass into and up the tower. The bait is dipped for a few seconds to let the glass bind and stick to the wire. The tower rollers were reversed and the bait drawn upwards. Two operators with flat nosed pinchers held and froze the edges to stop them drawing in. As soon as the sheet was well up the tower and the edges held, the edge bowls were fitted and the edges stabilised. The edge control panels are put on and sealed up with bull-muck. Bull-muck was a mixture of asbestos fibre and refractory clay that was universally used to seal low temperature joints and to stop the in-draught of air. Wing pads were fitted to the edge bowls to direct cooling air on to the edges. This cooling air was controlled by shutters in ducts incorporated in the control panel. Once the bait emerged from the top of the tower it was broken off. Thereafter, a continuous sheet of glass emerged vertically from the tower. As it emerged it went through a clamp which had two copper wires on each side the full width of the sheet. When the glass reached its full height the wires were clamped on to the glass and at the same time heated by an electric current so that the cut-off man could crack off the sheet with a drop of water and the carriers-off take the sheet to the cutting table. The thickness of the sheet of glass was controlled by the speed of draw and any variations in the thickness controlled by placing insulating pads on the coolers. A flat glass drawing machine would work for about 4–5 weeks. The problem was that devitrification formed under the edge bowls where the glass was drawn off the face of the breast block starving the feed of glass to the edges, with the result that the machine had to be put down to melt out the devit.

The Progressive Development of the Flat Drawn Tanks and Machines from 1930 to 1980

The first mechanical advance was in about 1934 when the vitrea cutter was introduced to replace the hot wire clamp. The vitrea cutter was a cutting wheel on a "rabbit" which ran across the sheet on an inclined plane to offset the speed of draw. The wheel scored the glass so that the carriers off could break off the sheet and put it on the cutting table cut off the edges and cut up the sheet into stock and cut sizes.

From 1936 to 1939, the speed of draw steadily increased from 30" per minute for 24 oz to 36" per minute. with corresponding increases for the other thicknesses. There were problems with lines and some seed, knots, blisters, and drops. The glass made was graded into various qualities, when it was bad for lines and ream, it was graded horticultural, ordinary window glass was graded as OQ, and a slightly better quality graded SQ. The best glass for seed and lines was graded SSQ or Triplex quality. The aim was always to make the best quality but there were uncontrolled variations. Reductions in the amount of seed in the glass were made by the introduction of fusion cast side block (Corhart) in the sides of the working end of the tanks. In the glass drawing towers there was a problem with the glass being scratched by the asbestos rollers. In the early towers the rollers were driven by a vertical shaft and open bevel gears. Some improvement was made when the bevel gears were replaced by worm wheels. As the speeds increased it soon became obvious that the edge bowls were unable to keep pace for the edges could not be drawn fast enough and the sheets were narrowing. The introduction of water cooled edge rolls was the complete answer. The edge rollers were driven off the main drive shaft on the tower and their speed was therefore automatically regulated to the speed of the draw. The edge fork beneath the rollers kept the edge in place. The effect of the edge rollers was to reduce the size of the shoulder ahead of the edge and so reduce the edge loss from 15% to 12%.

1939 saw the outbreak of the war with Germany so that in the period up to 1944–45 all the effort was concentrated on maintaining production. A visit in 1945–46 to America to see the sheet glass making plants showed that whilst the company had to stand still they had been making progress. They had been able to continue using almost pure white sand and natural gas whereas our importation of white sand from Belgium had ceased and we were using local sands of varying quality. Iron was endemic in our sand whilst in America they had to pill the sand, which is another word for adding a controlled amount of iron oxide and anthracite to the batch. They did not use producer gas for their furnaces were almost entirely fired with natural gas. Window glass making over there was easy compared to making glass in St. Helens. However, they had made one great advance and that was with blanket filling. It was the creation of the suspended end wall that enabled a very wide pocket to be built so that instead of filling the tanks with lumps of frit and cullet the batch was fed on to the surface of the metal in the form of a continuous blanket, over almost the full width of the tank.

After the visit to America, the company tried a very simple blanket feeder on one of the small Rolled Plate Tanks using a very wide filling pocket arch designed by Gaskell. The whole set-up worked very well and it was decided to equip the sheet tanks with suspended end walls and sophisticated blanket feeders. These feeders enabled more glass to be melted to keep pace with the steadily increasing drawing speeds. Upstairs on the top cutting floors where the glass sheets were cut off it was becoming more and more difficult to keep pace with the drawing machines. Here the first step was to do away with carrying off by hand and to carry off the sheet of glass using suction pads on a lever arm suspended from a butcher rail. This in turn allowed edge cutters to be used on the machine instead of on a cutting table.

Speeds on 24 oz went up to 45″ and 50″ per minute with the result that the annealing of the glass was not all that it should be. So far improvements in the control of tower temperatures had kept pace with annealing, but it had got to the point where it was necessary to establish more precise control of the annealing temperatures throughout the height of the tower. This was done after a great deal of experimental work using hot gas injection.

Whilst all the developments were going on the demand was such that an additional sheet No. 8 tank had to be built making use of better refractories and regenerator bricks. For many years the water boxes were fed with water from the canal for towns water was too expensive. This water was filtered through elementary sand filters, and pumped through the water boxes, then sprayed into the canal to cool it before it was re-circulated. Even this did not stop elongated frogs from being found in blocked coolers, nor did it stop sand and dirt from getting into the system. Blocked water boxes would burst and leak water into the tank. Replacing them was a regular job for the tank menders who had to work on the tank sides under the stacks in very hot conditions. It was the tank menders who kept the tanks going, often facing worn side blocks with tank side water boxes whilst the tank was still working. A closed circuit water cooling system in which the water was chemically treated and circulated through the water boxes and cooled in large air cooled radiators was tried out. It was expensive to run, but in the long term, was very successful in the benefits to water box life, tank life and human effort.

Up to now all the tank had their own mixing room, so the next major step was to build a central mixing room on the site of the derelict No. 8 tank to serve the whole of the works. This mixing room was designed to be completely automatic and to supply all the sheet and rolled plate tanks. It successfully eliminated some of the human error and by delivering batches of frit with each ingredient accurately weighed. However, if and when it did go wrong, it went wrong in such a big way that it was easily detectable. Although it was a large capital investment the savings were very great indeed.

Platinum coated drawbars were introduced to reduce bubble in the sheet and platinum coated floaters to reduce wear. 1950 saw the expansion of sheet glass making plants in the Commonwealth and the greater use of mechanical producers.

The early experimental work on the float process was started in Sheet Works and from 1958 the expansion of the float process throughout the world.

The Horseshoe Fired Tank

Although nearly all tanks making window glass were conventionally fired across the melt end Siemens did file a patent for a horseshoe fired tank. This kind of firing was very suitable for melting glass for making bottles and cast glasses. It was used by Chance Brothers for making rolled plate glass up to and into the 1950s.

CHAPTER 12.

Materials for Making Glass

At the beginning of the research on window glass making at Grove Street the composition or the recipe for the mixture of raw materials used for glass making at Sheet Works from 1826 to 1900 was quite unknown. As the work progressed it was found that there were no other references to materials in the records other than sand, salt or nitre cake, and limestone. The sand was either local or bought in from abroad. There were many references to salt and nitre cake, some being bought in and some being made in the works. Limestone was obtained from various sources and some of it from the quarry in Hard Lane, St. Helens, the remains of which, can still be seen. It therefore soon became clear that it was a simple silica, soda, lime glass that was being melted.

Little was known about the kind of glass, nor its precise composition, that the company made from 1826 to 1863 but it can be safely assumed that it was a silica, lime, and soda glass. No doubt that owing to the variation in the quality of the raw materials there would have been some variation in colour. From 1863 we get some indication for in 1867 arsenic was introduced. Arsenic is a decolouriser and it may have been necessary to reduce the colour brought about by the variations in the iron content in the sand. There were many references to dirty sand and the inadequacies of sand washing.

In January 1871 the company was making blown sheet glass from pots and there was a note about sheet mixtures.

	lbs	lbs	lbs
Dry Sand	760	800	800
Saltcake	322	300	326
Limestone	319	250	170
Charcoal.	24⅓		
or Anthracite	16	8	16
Arsenic	6	6	6
	Blinkhornes	Old Standard	Rolled Soft
	1431	1364	

The references to Blinkhorne, Old Standard and Rolled Soft are as yet unknown.

In Rolled Plate, Board. Min 396, stated "made an alteration to soften it reducing the limestone to 60 lb per batch of 800 lb of sand, this in one pot and the result very satisfactory".

In 1920 the composition of machine blown cylinder and hand blown sheet glass was given as:-

SiO_2	71.73%
CaO	13.15%
Na_2O	11.13%
Al_2O_3	0.72%

The following table gives an interesting account of the compositions from 1923 to 1943:-

		SiO_2	CaO	MgO	R_2O_3	SO_3	Na_2O
Hand blown sheet	1923	72.7	13.5	1.9	0.51		1.4
Cylinder drawn sheet	1933	72.6	13.3	1.8	0.5		11.8
PPG	1933	72.5	8.9	2.7	1.91	0.5	13.54
Flat Drawn	1942	72.6	8.9	2.8	1.6	0.5	13.64
No. 6 modified	1944	72.1	8.8	3.4	1.6	0.5	13.6

Despite what was given in the table above George McOnie in his diary for No. 6 Tank Machine Blown Cylinder Glass, gave the composition as follows,

SiO_2	73.1%
CaO	14.3%
Al_2O+Fe.	1.1%
SO_3	0.7%
Mg.O	0.3%
Na_2O.	10.5%

The batch in No. 6 now contained 90% soda ash and 10% saltcake. In the early days the use of saltcake only in the batch gave rise to sulphur, and to large amounts of free salt which attacked the refractories. The introduction of Soda Ash in No. 6 Tank allowed the glass maker to reduce the amount of saltcake with benefits all round.

The glass maker strives to make a glass that is economic to melt and at the same time marketable for weathering, cutting and glare.

The glass made on the Fourcault process at Queenborough was softer than that made on the PPG process in St. Helens and care had to be taken to avoid glass stocked in piles getting wet, otherwise it devitrified. Although it was quite satisfactory as a window glass it did not weather quite as well as the St. Helens glass. In 1927 the Company made "Vita" glass on No. 14 Tank at Sheet Works. This was a clear white glass, made either from white sand or by the use of a decolouriser. When glazed, it let through the ultra violet causing all the colour to fade in the curtains and carpets. The green coloration brought about by the iron in ordinary window glass reduced the amount of the ultra violet let through.

Colour and composition had its effect on the fluidity of the metal and on tank wear. When the company made spectacle glass in a rolled plate tank, the heat penetration and the circulation was such that the bottom blocks were red hot and the wear was such that when the tank was emptied, the bottom was only inches thick in places.

Other Glass Compositions and Recipes Used Elsewhere

In 1860 in Ure's *Dictionary of Arts* the composition of crown glass was given as :-

Silicic Acid	63%
Soda or Potash	22%
Lime	12%
Alumina	3%

and for green window glass

Silicic Acid	69%
Soda	11%
Lime	13%
Alumina	7%

In Bontemp (1868) there is a reference to the effect that the batch in general in use in Belgium and England was:-

White sand	100
Potassium Carb.	7
Lead Oxide	5
Sodium Carb.	50
Sodium Nitrate	5
Arsenic	0.5
Powdered chalk	11

In the illustrated itinerary by Cyrus Redding dated 1842, the proportion of materials necessary to produce good plate glass were:-

Siliceous sand	720 lb
Alkaline salt	450 lb
Quick lime	80 lb
Nitre	25 lb
Cullet	425 lb

Professor W. E. S. Turner in his paper to the Institution of Mechanical Engineers in 1930 reported on the developments in the composition of Fourcault glass. At first it was:-

SiO_2	71–73%
CaO	13–15.5%
Na_2O	11–13%
Al_2O_3	0.7–2%

As this was subject to devitrification in the drawing chamber and atmospheric corrosion it was modified to:-

SiO_2	72–73%
CaO	10–11.5%
MgO	1–2%
Na_2O	13.5–15%

Faults in Glass Making

There were many different faults that could occur in the making of window glass. There were those created whilst the glass is being made and those created by bad workmanship afterwards. Bubble could arise from incomplete fining or from claywork in the kiln. Ream in the glass was generally due to a heavier glass probably containing alumina drawn from the claywork. Drops were like tear drops in the glass which came from the underside of the rings and from the crown. Blisters on the surface of the glass were due to bubble on or near the surface. Salt blisters were white in colour. Other faults were cockle and scratches which occurred whilst the glass was being flattened. Cockle was caused by the lagre and scratches came from either the polissoire or the fourchette.

Sulphur was one of the principal faults in the past. Sulphur was a deposit on the surface of the glass that was very difficult to remove.

It occurred when blowing over the metal in the tank, as well as in the blowing furnace and in the flattening kilns. Throughout the books of the company, apart from the period when crown glass was made, there was a continual reference to sulphur on the glass. At an early stage it was thought to be coming from the melt and attempts were made to cure it by adding sawdust to the batch. Sulphur was less evident with coal firing and more prevalent with gas. For years the process suffered from this fault. It came and went for no apparent reason. Trials were made with some success to purify the gas, but never was it attributed to sulphur in the coal. Sulphur became less when they began to use the coal from the Nottingham and Durham coalfields. It was finally cured in the machine cylinder and flat drawn glass processes by sealing off the drawing kilns completely from the tank and firing them with town gas.

CHAPTER 13.

Historical Summary of 150 Years of Sand Getting

The Era of the Horse and Cart—1826 to 1881

This is a map showing what was thought to be the roads around

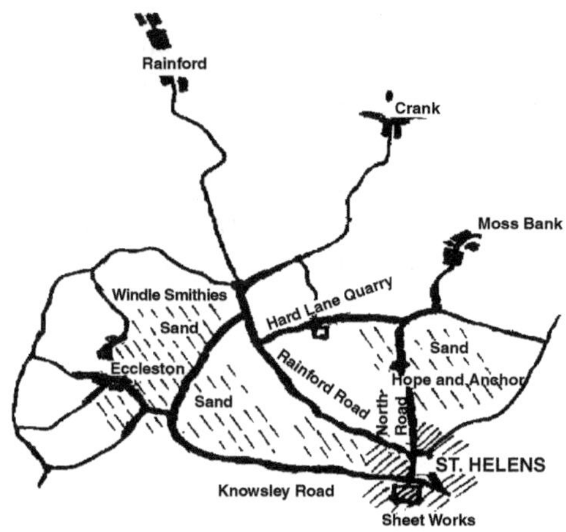

St. Helens from 1826 before the railway to Rainford was built in 1851. The country roads at that time would most likely be un-metalled and paved with limestone, whereas the roads in St. Helens were cobblestoned. Although James Taylor in the Liverpool essays on "The Shirdley Hill Sands" in 1967 suggested a concentration of sand working in the Skelmersdale area and areas well to the north of Rainford, there was no indication in the Pilkington Archives that the company carted sand from so far afield. All the indications were that besides importing sand the company was getting it locally in the area around Eccleston , Windle and Haresfinch by cart down Knowsley Road, Rainford Road and City cum North Road to Sheet Works.

When the railway became operational in 1851, sand could be carted to the rail heads at Rainford, Crank and Moss Bank. Up to 1873 horse drawn tram ways were being developed and later possibly between 1871 and 1881 haulage was more by a continuous wire rope, but it did enable sand to be gotten from a wider afield, possibly Rainford,

Knowsley and Crank, so that in 1879 it was proposed to lay down a tramway 2 miles long to Crank Station. It does however appear that during this period sand was gotten by a combination of tram ways, railways, and horse drawn carting, but it is still not known exactly how it all operated. There were indications that instead of washing sand in the brooks or in the works more local portable washing plants were being used.

The Era of the Tramway—1881 to 1930

With the help of Messrs Townley and Peden, the authors of books on railway systems, who had access to and identified the tramways on earlier survey maps of the area, it would appear that between 1873 and 1881 there was a steady build up of experience in the development of tramways and the use of wire rope haulage, which resulted in a decision to work the sandfield to the south of the main line and to install a tramway system complete with a tippler, a new sandwash plant, and a siding at Mill Lane in 1881. The use and development of the Rainford Rookery and Mill Lane fields was in line with the findings of James Taylor in his Essay.

The map shows the area to the south of the main line at Mill Lane. In this area there were almost permanent tram ways laid to the Rookery area at Rainford and south to the Cricket Field and the areas north of the Bottle and Glass public house. It was interesting to note that there was a clay pit behind the Bottle and Glass which also served the Pottery at Mill Lane.

Although this area was the main source of sand supply there were indications that sand was still obtained from other local areas as well as from abroad. In 1914 the sand washing plant was set up at Washway lane for sand from the Crank area. By 1930 sand getting had extended to Kings Moss and Billinge still using tramways but now equipped with tipping bogies and mechanical excavators.

The Era of the Ropeway—1930 to 1961

By 1930 sand getting operations were concentrated at Kings Moss using a ropeway to convey the sand to the washing plant at Mill Lane. The head of the ropeway, not shown on the drawing, was serviced by tramways that were moved from field to field.

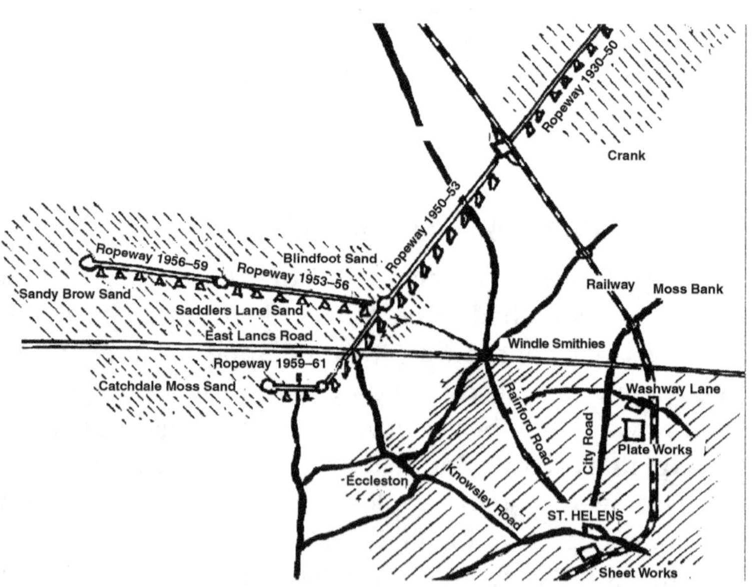

Mechanical excavator and tramway

Later as shown the ropeway was moved to Saddlers lane, then by installing a transfer station or angle station the ropeway extended to get the sand from Blindfoot. When those fields were exhausted, it was extended to Sandy Brow. Finally, the transfer station was moved to Catchdale Moss.

The Era of Road Transport—1961 to 1978

In 1961 the ropeway was dismantled in favour of road transport and the need to obtain sand from even further afield, in places such as Skelmersdale, Bickerstaffe and Ormskirk.

The ropeway

CHAPTER 14.

Glossary—A Guide to Glassmakers' Terminology

The object of this guide is to explain the nature of the plant and the tools. For instance glass makers always talk of a "tank". **A tank is a glass melting furnace**. There are different types of tanks and they vary in size. The different types are described by their function. A "Flat Drawn Tank" is one for melting and making flat window glass. Other tanks are for melting and making bottles, plate glass, float glass, rolled plate glass, spectacle glass, and television tubes, to quote just a few.

The first tank that was made for melting glass was about 15 feet long, holding perhaps about 10 to 15 tons of glass and it was sometimes referred to as a **'Cistern or Cuve'**. When it was empty it would have held about **two mini cars**, side by side. After it was enlarged for melting and making blown sheet glass it would be about 45 feet in length and would have held **4 family cars parked end to end with room to spare on either side**. As time went on tanks were increased in size and capacity to 80 feet in length. They then would have held about 150 tons of glass and would have accommodated **14 family cars, that is two cars side by side and seven cars end to end**. Eventually much larger tanks were built, 100′ to 120′ in length, for melting and making machine drawn cylinder glass and later, flat glass. These tanks held between 800 to 1200 tons of glass and would have been able to hold **32 large cars, or 4 cars side by side and 8 cars end to end**.

A Pot Glass Melting Furnace was coal or gas fired to a temperature hot enough to melt and make glass in pots. The glass was made from sand, ground up limestone, and saltcake which was mixed together to form a **batch**, generally referred to as **frit**, which was loaded into **pots**. A pot was made from clay somewhat like a large flower pot in shape. It was burnt hard in a **pot-arch, burning kiln or ring arch**. Inside the pot was a burnt clay **ring** that floated on the surface of the molten glass generally referred to as **metal**. The pots were loaded into the pot furnace on to shelves. These shelves were called **sieges**.

When all the pots had been emptied they were refilled and the whole process of filling and emptying was called a **found**. There would be about 3–4 founds in any one week.

The molten glass in the pot was **gathered** from inside the ring on the end of a **blow pipe by the glassblower**. The blown glass was taken off the blowpipe by a **pontil or punty**, by the **punty sticker**. A pontil was a long iron bar about ¾″ in diameter with a blob of molten glass on the end. When blowing glass to form of a cylinder, the pontil was replaced by a **cordeline**. A cordeline is a short iron bar about ½″ in diameter with a crucifix handle. It was used for gathering a blob of molten glass to be put on the end of the cylinder and for winding a thread of hot glass round the cylinder so that it could be cracked off the blowpipe. Blown cylinders of glass, that were to be cracked off the blowpipes were placed on a trestle called a **chevallier**. Finished cylinders were taken to the cutting rooms for splitting by the **splitter**. The very large cylinders made by machine drawing were so large that they were split into two halves on the chevallier. These halves were called **shawls**. Broken glass called **cullet** was collected for re-melting Very small pieces of glass left over when making crown glass were cut into small window panes called **quarries**.

Cylinders of glass had to be split and then flattened. This was done by a **flattener** in a **flattening kiln**, assisted by a **pusher**. The flattener used a **crappe** to lift the cylinders from the **dollum** on to the **flattening stone**. The crappe was a long iron bar with a handle. The dollum was a turntable for warming the cylinders before they went into the flattening chamber. A flattening stone was a flat slab of burnt clay about 4″ thick and up to 72″ long by 42″ wide, with a piece of thick glass called a **lagre** on top of it. The flattener used the crappe to open up the cylinder as it softened in the heat of the kiln in such a way that it would lay flat on the lagre. He finished off the flattening process with a **pollissoire**. A pollissoire was a block of wood on the end of a long iron bar. The iron bar was pointed at one end and the blocks of wood had a hole down the centre so that they could be spiked on to it. After the cylinder had been flattened, the flattener took off the flattened sheet of glass with a **fourchette** and placed it in the **lehr**. The fourchette was a long iron bar with flattened tines on the end.

These tines were very sharp. The pusher pushed the stones around and operated the lehr. A **lehr** is a long tunnel down which sheets of glass were slowly cooled in order to anneal them. At the end of the lehr the glass was dipped in a **dipper** to clean it of **hum**. Hum is a word used to describe a smoky film that was deposited on the surface of the glass in the lehr. Glass was annealed by being cooled down slowly in a **piling kiln** before lehrs came into use. It was put into the kiln by a **piler** and leaned against a **drosser**.

In the process of making crown glass the blowers used a **nosing kiln, a bottoming kiln and a flashing kiln** to shape the gather and to flash it out into a **table.** A table is a round sheet of glass with a punty mark or **bullion** in the middle. The kilns were merely sources of heat to keep the glass sufficiently soft for it to be worked

In the process of making blown cylinder glass it was the custom to use the glass furnace pot gathering holes as a source of heat. This was known as **"blowing over the pots"**. This slowed up the whole process of gathering and to speed it up **blowing furnaces** were built. A blowing furnace or **glory hole** was merely a source of heat to keep the glass hot so that it could be blown. Blowing furnaces were in turn done away with when it became possible to gather and blow in a tank. This was known as **"blowing over the metal"**.

A Tank or Glass Melting Furnace was a gas fired **regenerative** furnace. In this type of furnace the gas and the air were switched over from side to side every 20 minutes so that they could be preheated in the **regenerators** by the out going hot gases. In the regenerators was the **stuffing** supported on **saddle tile**. The tank holding the metal was constructed of **tank blocks** supported on **channel tile**. It was covered by a superstructure consisting of **side walls** and a **crown** made of **silica brick**. The gas was supplied by a **gas producer** operated by **producer men**. The men who kept the tank **filling pocket** filled with frit and cullet were called **teazers**. They used **lump hooks** to push out and control the lumps of frit floating on the metal. Further down the tank the **skimmers** using **rakes and ladles** skimmed off any debris that floated down the tank on the surface of the metal. The gas and the air were fed into the tank through the **upcasts** from the regenerators, into the **stacks**, and then to the burner **ports**, at what was called the

melt end. The gas and air were separated by the **mid-feather in the ports**. There were small **sight holes** in the side of the tank to view the **flame length** and **the rate of melt**. The balance of the pressures in the tank was indicated by the **flush** from the sight holes. Sight holes were closed by a brick or a **stopper tile**. A stopper tile was made of burnt clay about 12″ square about 1″ thick with a 2″ hole in the middle so that it can be lifted into place on **stopper rod**. In the middle of the tank was a **skimming pocket**. This pocket was used by the skimmer and it was also used for **ringing the tank**. This expression was used when it became necessary to replace the gathering rings with new ones and the same expression was used to this day when **floaters** were put in. Skimming pockets were closed by **tweels**. A tweel was a shutter made up of clay bricks in a framework of iron. It was on a chain with a balance weight so that could be moved up or down to close an opening in a furnace or kiln. Floaters were pieces of burnt clay that float on the surface of the metal and were there to steer any debris into the skimming pocket. Beyond the melt end and the floaters there was what was called the **working or gathering end**. In the working end of a tank making blown cylinder glass there were **gathering** and **blowing holes** as well as a **pipe hole**. A pipe hole was there to heat the blowpipes so that the metal could be gathered.

In a tank making flat glass the flat glass drawing machines were located above a **bath**. The bath was sealed off from the working end of the tank by a **shut-off**. There was a **skimbar** in the bath to clean the surface of the metal and **draw bar** beneath the metal to stabilise the sheet of glass as it was drawn upwards into the **tower**. The drawing kiln was covered by a **front arch, and a curtain arch**. The drawing kiln consisted of two **ell blocks, water coolers and pads, edge bowls or edge rollers**. The kiln arches, ell blocks, and end walls were sealed of with **bull muck.** Bull muck was a mixture of asbestos fibres and wet clay.

Light clothing was worn complemented by woollen or felt **mitts** for handling the tools when working in the tanks. For really hot work for tank repairs or ring changes, it was usual to wear **gladdins and asbestos mitts**. Gladdins were a set of trousers and jacket made in felt with a felt hat and cover for the back of the neck.

When the metal was bad for seed or ream, it was the custom, mostly at the weekends when there was no load on the tank, to heat it up or **'found'** it. During the founding, the metal was stirred up by **plugging the tank** with a potato or a piece of wood impaled on the end of an iron rod. Any root vegetable would have done, even wet brown paper rolled in a ball stuck together with paste. The potato or plug on the end of the rod was pushed down into the molten metal to the bottom of the tank. The steam and the gases that were given off created quite an upheaval and had the effect stirring up the metal quite considerably.

Sometimes it was necessary to empty a tank by running off the metal. This would have to be done if there was to be a change in the composition of the glass. It was possible to empty a 350/400 ton tank by running off the metal down below into the cave, where the glass would solidify and be chipped out later for use as cullet. Emptying a 1200 ton tank was a different matter for it involved such a large quantity of glass. No. 6 cylinder and sheet drawn glass tank was a case in point. This tank was built in such a way that the molten glass could be run off into a compound. Even then there was still too much glass to allow it to solidify, so the molten glass was run off down a chute and **crizzled** in a torrent of water. Water sprays.in the compound all helped to **crizzle** the glass into small particles so that it could be more easily handled.

Parging up a joint or an opening was to fill or grout it with a clay mortar or **bull muck**. There were expansion joints in the silica work that never completely closed up and there were joints around the drawing machines that needed to be sealed or plastered over.

ACKNOWLEDGEMENTS and REFERENCES

All the background information for the record of window glass making in the Company at Sheet Works, unless said otherwise, has been obtained from the Board Minutes of the Company from 1863 to 1940 and other records held in the Pilkington Archives. This record could not have been completed without reference to other sources and help from elsewhere. These other sources together with acknowledgements are as follows :-

"Le Verre et la Cristal" published in 1897 by M. J. Henrivaux sent to me by D. M. van Gogswaardt of Amsterdam. Early history of glass making and the beehive furnace.

"Guide du Verrier". Traite Historique et Practique de la Fabrication des Verres, Cristaux, Vitraux. by G. Bontemps 1868.

"Now Thus— Now Thus" 1826 to 1926.

"The Glassmakers" by Professor T. C. Barker, particularly for the works plans.

"An Illustrated Itinerary of the County of Lancaster" Cyrus Redding 1842.

"Machinery and Methods of Manufacture of Sheet Glass" Professor W. E. S. Turner.

"The Manufacture of Blown Window Glass" by Henry Deacon F.C.S. from whose paper I am indebted for the illustrations of the blowing of sheet and crown glass.

Richard Pilkington. Pot melting and blowing 1864. L405.

The Reminiscences of Ernie Baddeley.

Reminiscences Vol. 1 and Vol. 2.

Tank Repair Books and Tank Drawing Books.

William Baynton's Diary October 1790 on his visit to the Newcastle Glass Bottle Houses.

George McOnie's Diary April 1927 PB 536.

The valuation sheets and map of the works of 1878 PB 273.

Lancaster University Archaeological Unit. Mick Krupa and Jamie Quartermain for their co-operation and work on the excavation of No. 9 Tank

Harry Langtree T.D. block minder and gatherer at No. 12 Tank whose contribution about gathering and blowing was invaluable and whose drawings, illustrating many facets of the process have been edited and recorded.

Jimmy Hill who started work as a time boy in the building department in 1919 at the age of 14, who went all over sheet works in the early 1920s and saw all the tanks, producers, and flattening kilns, especially the workings under the Old Cone. A furnace builder all his life and an expert on glass furnace construction, responsible for building furnaces all over the world.

Arthur Reeves of Chance Brothers, an expert on the design of blown glass furnaces and the inclined grate gas producer. A glass maker and historian of glass making at that company.

Evan Davies whose uncle was a flattener, started work as a pusher in 1919 at 8 s per week rising to 12 s. Became a fully qualified flattener at 21, went on to drawn cylinder flattening in 1925, then on to be a chief operator in flat drawn.

Danny Hurst a splitter of blown glass cylinders in the 1920s, then on to top floor cutting in flat drawn and finally one of my top floor supervisors in that department.

Additional information and help from Harry Telford, a time gatherer and later machine man in flat drawn, and Eric Hiorns, an electrician who knew of the underground tunnels and who helped me in the use of an Amstrad Computer.

More help from Edwin Harrison, Jack Houghton, Frank Lockwood, Jack Clitheroe, Bert Jones and E. M. S. Wood. Little bits of local history from Arthur Huyton's essays, Mike Hevey, Norman Bowdler of Bowdlers Ironmongers formerly in Bridge Street, and Tyrer's of Bridge Street.

Jim Winstanley who wrote up all the notes and information on sand getting from 1946 to its end in 1980 when all the records were destroyed as of no further use. To John Virgoe for all the copies of his notes on sand quality and sand getting, including the extensive and detailed report on sandfield operations made by S. R. Glover. To Rex Lowe and the staff at Lathom for an insight into what records there were on the sandfields and for their interest in the past history of glass making.

I am indebted to Dinah Stobbs and to the Company Archives for the accommodation and the constant supply of company records, papers and references without which this record of the glass making heritage of the company could not have been attempted let alone written. Grateful thanks to Dr David Martlew for his encouragement and help in the preparation of this monograph.

R. A. Parkin

Cutting from the *St Helens Star*, 2001:

AN adopted son of St Helens, who has contributed to the business and sporting life of the town over almost 50 years, has died aged 88.

Ronald Austin Parkin (pictured) passed away at his home in Hard Lane following a brief illness. He leaves his widow Anne, son John, daughter Rosemarie, two grandchildren and one great-grandchild.

Born in London, raised in Lincoln and a graduate of Glasgow University, chartered engineer Ron was a past member of several professional bodies and was employed by Pilkington throughout his working life.

He was formerly works manager at Triplex and the glass giants' Queenborough plant, and a member of the North West Water Authority, CBI production committee, joint metrication standards commitee, and Institution of Mechanical Engineers.

Mr Parkin's sporting involvement included presidential positions with St Helens Recreation hockey club, Ruskin Park RFC, Pilkington Recreation club, Windle Bowling Club, St Helens Table Tennis League, and membership of the Sports Council.

Sailing was Ron's other passion, for he also enjoyed similar roles with the Lowton and Pilkington clubs, the National Schools Sailing Association, he was commodore of Leigh sailing club and member of Manchester Cruising Club, and was on steering committees at a number of other clubs.

He also published a book entitled 'Window Glass Makers of St Helens' which is available at the Glass Museum.

A private funeral service will be held today, (Thursday, March 15).

● By Denis Whittle

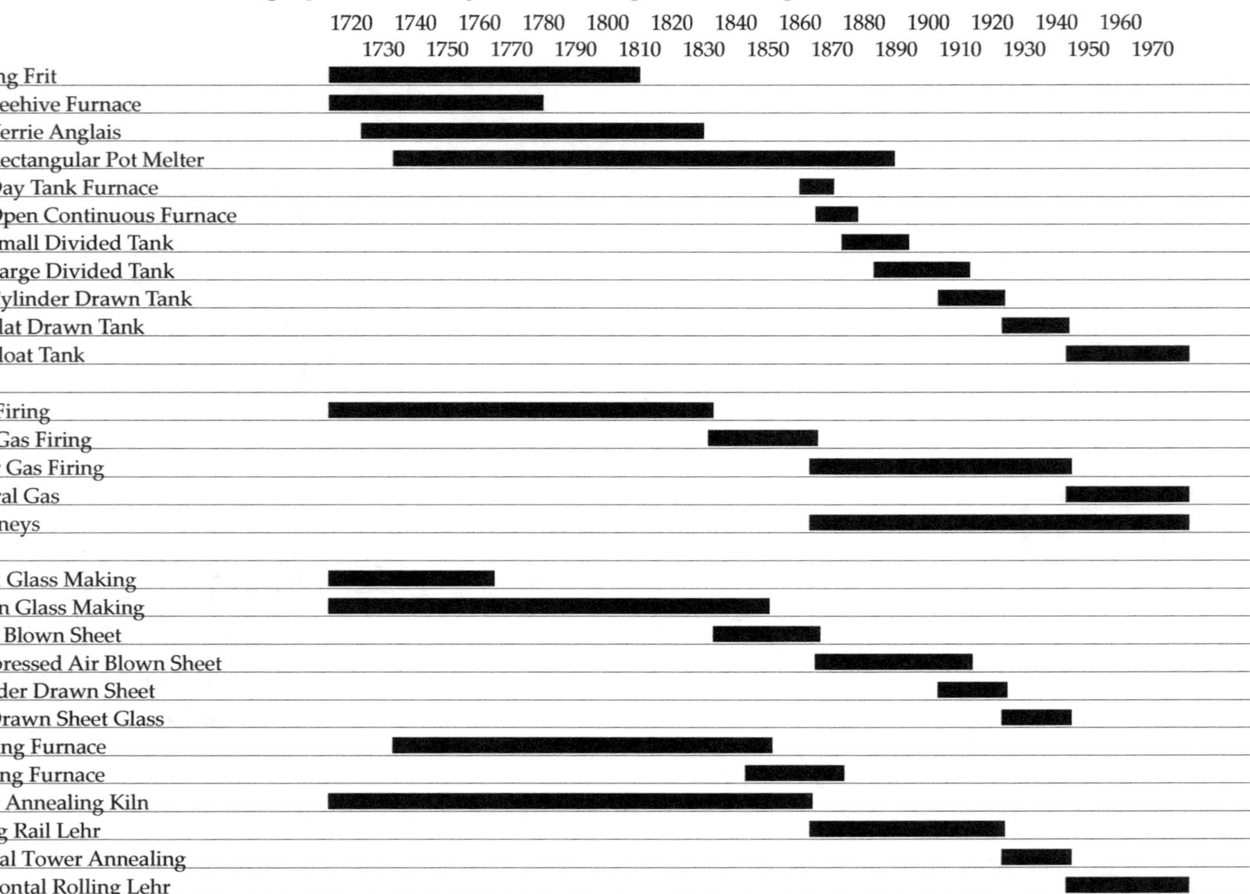

A graphical history of Window glass making from 1720 to 1970
1720 1730 1740 1750 1760 1770 1780 1790 1800 1810 1820 1830 1840 1850 1860 1870 1880 1890 1900 1910 1920 1930 1940 1950 1960 1970
Making Frit
The Beehive Furnace
The Verrie Anglais
The Rectangular Pot Melter
The Day Tank Furnace
The Open Continuous Furnace
The Small Divided Tank
The Large Divided Tank
The Cylinder Drawn Tank
The Flat Drawn Tank
The Float Tank
Coal Firing
Coal Gas Firing
Water Gas Firing
Natural Gas
Chimneys
Broad Glass Making
Crown Glass Making
Hand Blown Sheet
Compressed Air Blown Sheet
Cylinder Drawn Sheet
Flat Drawn Sheet Glass
Flashing Furnace
Blowing Furnace
Piling Annealing Kiln
Lifting Rail Lehr
Vertical Tower Annealing
Horizontal Rolling Lehr

www.ingramcontent.com/pod-product-compliance
Lightning Source LLC
LaVergne TN
LVHW020639100826
845148LV00012B/2250
9780900682971